Bernard Montgomery's

Art of War

"Author and military historian Zita Steele certainly did her homework in giving us 'Bernard Montgomery's Art of War.'

Not a biography, although the book lets the reader know quite a bit about the man, this is a well organized compilation of Montgomery's thoughts and writings on his approach to war. Montgomery served as a young British officer during World War I and then later as a top general in World War II. He was also a student of warfare, believing that to be really good at leading in war, one had to be an expert on past masters of the art.

Steele organized Montgomery's views and beliefs in ten chapters, each focused on a different topic. These topics include his core principles, approaches to battle, the spirit of the warrior, battle management, and more.

Montgomery's teachings should be mandatory reading material for all young officers, and this book can give you a quick primer. I recommend it."

—Bob Doerr, Military Writers Society of America

"This book brings Monty up with the great leaders of history, his contribution to the final Victory in WWII has never been really appreciated and this book rights that wrong. The leadership chapters should be strongly recommended reading for any future military leader (or politician) and military historian. As a former soldier I can relate to many of the leadership aspects mentioned in the book, the things I did right or wrong!"

—Major David Seeney, Retired, late Royal Warwickshire Regiment and Airborne Forces. Chairman of Friends of the Royal Regiment of Fusiliers Museum (Royal Warwickshire)

Bernard Montgomery's Art of War
Zita Steele
Bernard Law Montgomery

Fletcher & Co. Publishers
© May 2020, Fletcher & Co. Publishers LLC.

Author: Zita Steele
Cover design by Zita Steele, public domain photos.

Interior design: Noël-Marie Fletcher
Photo of Zita Steele: Noël-Marie Fletcher

All rights reserved, including the right to reproduce this book, or portions thereof, in any form without written permission except for the use of brief quotations embodied in critical articles and reviews.

Cataloging-in-Publication data for this book is available from the Library of Congress.

Library of Congress Control Number: 2020938650

Cataloging information

ISBN-13 978-1-941184-35-6

First Edition
Published in the United States of America

Bernard Montgomery's

Art of War

by

Zita Steele

&

Bernard Law Montgomery

Fletcher & Co. Publishers
www.fletcherpublishers.com

Contents

Acknowledgments — 12

About this Book — 13

Introduction: Montgomery and His Art of War — 18

Chapter 1: Core Principles — 55

Chapter 2: The Way of the Commander — 61

Chapter 3: The Spirit of the Warrior — 80

Chapter 4: Approaches to Battle — 119

Chapter 5: Essentials of the Fighting Machine — 126

Chapter 6: Battle Management — 141

Chapter 7: Attacking & Defending — 156

Chapter 8: Thoughts on Nuclear Warfare — 175

Chapter 9: Thoughts on Peace — 180

Chapter 10: The Study of Military History — 188

Montgomery's Reading List — 202

About the Author: Zita Steele — 207

Dedication

This book is dedicated to the great spirit of alliance between Great Britain and the United States of America, which led our nations together to save Europe from tyranny through their victory 75 years ago this May 2020.

There must be NO WITHDRAWAL anywhere, and of course NO SURRENDER.

—*Bernard Law Montgomery*

This phrase was written by General Bernard Law Montgomery in a personal message to his troops before the Battle of Medenine in North Africa in 1943. The original document is held at the Imperial War Museum.

British World War II photo of Bernard Law Montgomery.

Acknowledgments

I would like to extend a special thanks to the following people and organizations:

• Major David Seeney, Retired, Royal Warwickshire Regiment and Airborne Forces, and Chairman of the Friends of the Royal Regiment of Fusiliers Museum (Royal Warwickshire)

• The Friends of the Royal Regiment of Fusiliers Museum (Royal Warwickshire)

• The Royal Regiment of Fusiliers Museum (Royal Warwickshire) in Warwick, England

• Dr. Iwona Korga, Historian, the Jozef Pilsudski Institute of America in Brooklyn, N.Y.

• The United States Holocaust Memorial Museum in Washington, D.C.

About This Book

General Sir Bernard Law Montgomery, Commander in Chief, 21st Army Group, in February 1944 during a meeting in London of the Supreme Command, Allied Expeditionary Force during World War II. Photo courtesy of Wikimedia Commons.

This book is a collection of essential theories and sayings by Bernard Law Montgomery that formed the basis of his approach to the art and science of war.

I have had a lifelong enthusiasm for military history, which began in my early life and continued through my teenage years and university studies, and has formed the basis of my career as an adult. I have done research in person at numerous military and historical museums, and institutions across the United States as well as in Europe. I have also visited historic sites and many battlefields. My areas of interest focus on ancient military history (for example, campaigns of ancient Greece, Persia, and Rome) and modern warfare, particularly World War I and especially World War II.

Bernard Law Montgomery is my favorite military leader of all time. When I was first researching Montgomery, I was surprised to learn he was a prolific writer. I find his writings impressive on many levels. I have collected most of his published works. Sadly, I've found that the study of his writings has been largely overlooked and neglected.

Although Montgomery wrote extensively on his philosophy, the fundamentals of his art of war have never been assembled in a single comprehensive volume.

My goal in creating this book is to give Montgomery's keen insights the treatment I believe they properly deserve: to edit, arrange, and assemble them in a simple format ideal for study and reflection.

Painting of Sun Tzu (left) and a bamboo book of "Art of War" commissioned or written during the reign of the Qianlong Emperor (1733-1796). Photos courtesy of Wikimedia Commons. The book is in the collection of University of California, Riverside.

In creating this book, I was inspired by Sun Tzu's "Art of War" (circa 5th century B.C.) and Miyamoto Musashi's "Book of Five Rings" (circa 1645).

Scroll illustration of Miyamoto Musashi wielding two bokken (wooden swords) from a wood-block print by Utagawa Kuniyoshi (1798-1861), image courtesy of Wikimedia Commons.

This book consists of freely arranged statements of doctrine and principle, written by Montgomery, and organized in themes. As editor, I categorized and arranged the material; some subheadings were already created by Montgomery. All themes and sections have been drawn from Montgomery's theories.

By "freely arranged" statements, I mean that Montgomery's statements are structured loosely and freely to provoke thought, except where sequential organization is needed for clarity. I drew inspiration from Sun Tzu and Musashi in adopting this style.

I selected passages from Montgomery's writings, including:
- *"A History of Warfare"* (1968),

- "*The Memoirs of Field Marshal Montgomery*" (1958),
- "*Morale in Battle: Analysis*" (1946),
- "*High Command in War*" (1945),
- "*Some Notes on the Conduct of War and the Infantry Division in Battle*" (1944),
- "*The Armored Division in Battle*" (1944), and
- "*Some Notes on the Use of Air Power in Support of Land Operations and Direct Air Support*" (1944).

I also have included some statements he made during interviews and in speeches. For the sake of clarity, I made some editorial changes to simplify text that was overly complex or outdated. Photos are included to enhance the text. All images included in the book are public domain.

I have written an introduction to provide an overview of Montgomery and his characteristics as a military commander. I hope this will enlighten readers to better understand Montgomery's military philosophy.

I hope "Bernard Montgomery's Art of War" will be useful to military historians, soldiers, and anyone interested in learning about Montgomery's fighting style and military leadership.

I conclude my introduction with a reflection for readers taken from Theodore Roosevelt's 1910 speech, *"Citizenship in a Republic"*:

> *"It is not the critic who counts; not the man who points out how the strong man stumbles, or where the doer of deeds could have done them better. The credit belongs to the man who is actually in the arena, whose face is marred by dust and sweat and blood; who strives valiantly; who errs, and comes short again and again, because there is no effort without shortcoming; but who does actually strive to do the deeds.*

Who knows the great enthusiasms, the great devotions; who spends himself in a worthy cause; who at the best knows in the end the triumph of high achievement, and who at the worst, if he fails, at least fails while daring greatly, so that his place shall never be with those cold and timid souls who know neither victory nor defeat."

—Zita Steele

INTRODUCTION:
Montgomery and his Art of War

Bernard Montgomery as a Colonel. He served in many countries including India, Egypt and Palestine, and fought in three conflicts prior to World War II. Montgomery served actively in the military for 50 years. Public domain photo.

Bernard Law Montgomery (1887–1976) was a British soldier, writer, and military strategist most well known for his achievements as a successful commander. Montgomery rose to military prominence for numerous acts of bravery and intelligence throughout many conflicts beginning with the First World War. He later became famous as a General and Field Marshal during World War II.

Montgomery had a uniquely global perspective. Half Irish and English, he grew up in Australia, and later lived and worked in various locations in the United Kingdom. Throughout his military service, he traveled to many diverse countries in the world including India, Egypt, Palestine, France, Germany, Switzerland, Denmark, the United States, China, and Russia. Montgomery was fluent in French, Hindustani and Urdu.

Montgomery was a talented athlete and energetic British Army infantryman. He was captain of several rugby and football teams throughout his life. His teams were successful and won numerous trophies.

He joined the Royal Warwickshire Regiment and had the regimental symbol, an antelope, tattooed on his arm. He served actively in the military for 50 years.

A gifted and prolific writer, Montgomery began his writing career as a journalist for his regimental newsletter as a young man and contributed columns to it for many years. He authored more than seven books in addition to articles, speeches, and numerous publications. His work was published by *The Times* newspaper of London. Montgomery's writing interests included politics, literature, history, and military theory.

Montgomery saw combat in the field in World War I, the Irish War of Independence, a violent insurgency in Palestine, and World War II.

> Captain Bernard Law Montgomery, The Royal Warwickshire Regiment.
>
> Conspicuous gallant leading on 13th October, when he turned the enemy out of their trenches with the bayonet. He was severely wounded.

The *London Gazette* on Dec. 1, 1914 noted Montgomery's gallantry in World War I. With the Royal Warwickshire Regiment, Montgomery was part of the advance guard ordered to attack the Germans at Meteren in northern France. As his group advanced, the Germans fled into the village. "The Royal Warwickshire were then ordered to drive them out, and by skillful use of the ground made steady progress... Then they were halted whilst the attack developed in other quarters. This delay deprived the Royal Warwickshire of the credit of the actual capture of Meteren...and most of the casualties occurred whilst halted under heavy fire," according to *"The Story of the Royal Warwickshire Regiment"* by Charles Lethbridge Kingsford (1920).

During World War I, he was awarded the Distinguished Service Order for heroism on the battlefield. He was also made an Honorary Corporal of the 11th Battalion of the Chasseurs Alpins (Alpine Hunters), France's elite mountain infantry. He was wounded during action by gunshots to his lung and knee. He was also wounded in 1945 and lost most of his personal possessions to enemy bombings during World War II.

As a young man, Montgomery was noted by his superiors for his strong abilities as an analyst. He began the war in 1914 as a Lieutenant leading a platoon of 30 men. Within only four years, he had become the Chief of Staff of a division by age 30. He contributed to and developed many military training manuals based on his practical observations.

Montgomery with King George VI in the Netherlands in October 1944. Public domain photo.

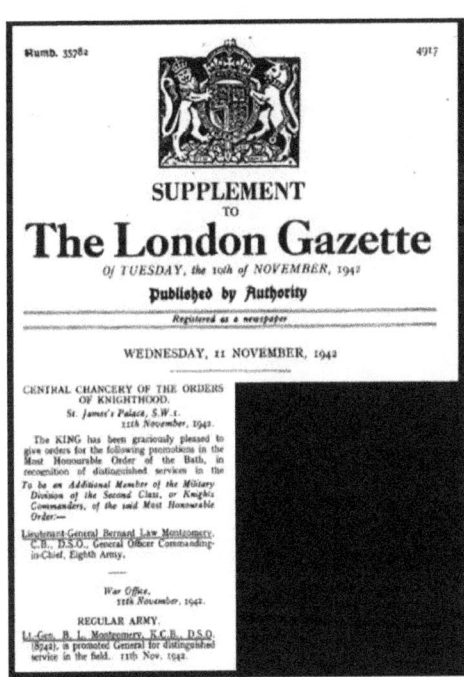

In 1942, King George VI promoted Montgomery to the rank of Knight Commander in the Order of the Bath, an order of chivalry founded in 1725. Two years earlier, the king had appointed him to the order's Military Division of the Third Class for notable service. Also in 1942, Montgomery was promoted to General for distinguished service in the field.

Montgomery's Battles in World War II

World War II Battle of France
- Battle of Dunkirk
- Dunkirk evacuation

North Africa Campaign
- Battle of Alam el Halfa
- Second Battle of El Alamein
- Battle of El Agheila

Tunisia Campaign
- Battle of Medenine
- Battle of the Mareth Line

Sicily Campaign

Italy Campaign

Western Front
- Operation Overlord
- Battle for Caen
- Operation Goodwood
- Operation Cobra
- Battle of the Falaise Pocket
- Siegfried Line Campaign
- Operation Market Garden
- Clearing the Channel Coast
- Battle of the Bulge
- Operation Veritable

Germany Invasion
- Operation Varsity
- Operation Plunder
- Battle of the Ruhr Pocket
- Battle of Hamburg

Montgomery became famous in World War II for his unexpected victories against Erwin Rommel, one of history's most aggressive generals, in North Africa. Whereas other noteworthy British commanders had failed to repel Rommel from his position threatening British control of the Nile, Montgomery succeeded in driving Rommel's army out of North Africa at a lightning pace. During this time, Montgomery showcased his unique style of fierce yet methodical fighting.

Montgomery's Eighth Army never lost a battle under his command nor was ever forced to retreat. Montgomery later commanded a total of 2 million men drawn from many nations after the Allied Normandy landings on D-Day in 1944. He was known for his humane and personal style of leadership and was well-loved by his troops.

Montgomery meets with King Christian X of Denmark after leading the liberation of Denmark in May 1945. The nation had previously been occupied by Germany since December 1939. Photo courtesy of the Museum of Danish Resistance Photo Archives.

DECORATION CONFERRED BY HIS MAJESTY THE KING OF DENMARK.

Order of the Elephant.

Field-Marshal Sir Bernard Law MONTGOMERY, G.C.B., D.S.O. (8742), late The Royal Warwickshire Regiment, Colonel Commandant

On July 5, 1945, King Christian X made Montgomery a Knight with the Order of the Elephant (Ridder af Elefantordenen), Denmark's highest-ranked honor and order of chivalry dating from 1693. This honor is typically reserved for royalty or heads of state. The photo above of the Order of the Elephant badge and medallion are courtesy of Wikimedia Commons. The notice above appeared in *The London Gazette* in July 1945.

Montgomery and the Royal Warwickshire Regiment

Montgomery joined the Royal Warwickshire Regiment in 1908 and served with the 1st Battalion for many years. Montgomery chose this regiment because he liked its cap badge insignia and also because it had a regular battalion stationed in India; Montgomery was extremely poor during his military school days at the Royal Military Academy Sandhurst and sought to obtain a posting in India to support himself on his own wages after graduation.

The regimental system is very important in the British military. British officers are commissioned into specific regiments and treated differently than other ranks. All British officers are regimental officers until reaching the rank of Brigadier; from that rank and above, they are General Officers. The regimental identity, therefore, is something that stays with a British officer for life.

The Royal Warwickshire Regiment has an illustrious military history. It was designated as the 6th oldest infantry regiment in the British Army. The regiment's early origins began in 1674 when it was formed by Prince William of Orange to defend Holland from the French. The regiment accompanied him to England in 1688 during the Glorious Revolution. The regiment also had fought during the War of Spanish Succession between 1702 and 1712, the Peninsula Wars in Spain and France from 1808–1814, the Kaffir Wars in South Africa from 1846–56, the Indian Mutiny of 1857, the Sudan campaign in 1898 and the Boer War from 1899–1902, and in other conflicts. The regiment also fought in North America during the War of 1812 and received the regimental battle honor "Niagara" for its distinguished valor in repulsing an American sortie during the Battle of Niagara Falls on September 17, 1814. King William IV bestowed the title of "Royal" on the regiment in 1832. It became known as the Royal Warwickshire Regiment in 1881. Men of the regiment are often called "Warwicks".

The regimental badge is an antelope. According to tradition, the antelope symbol was adapted from a Moorish banner captured during the Battle of Saragossa. However, the antelope also was used as a symbol by England's Lancastrian Kings and could have been used due to royal significance.

The regiment's mascot was an Indian Black Buck antelope called Bobby. Bobby the Antelope was a strong part of the regiment's identity; the regiment kept live mascots for 200 years. The first mascot, adopted by the regiment in India between 1825 and 1841, was named "Bobby." The 1st Battalion subsequently bestowed this name on all his successors.

The regimental mascot, Bobby the Antelope, inspects a drum while accompanied by his entourage prior to 1940. Photo courtesy of the U.K. Fusiliers Association.

Every Bobby looked very grand on parade; he wore a ceremonial outfit consisting of a small, embroidered coat and removable silver tips on his horns, and was escorted by white leashes. Off duty, each Bobby led a pampered life under the care of two mascot handlers. The antelopes lived in roomy enclosures, got daily walks, and enjoyed eating various treats such as biscuits, vegetables, sugar—and sometimes even cigarettes. Baby Bobbies, adopted from zoos, required bottle-feeding. The antelopes tended to be curious and were known to chew on various objects—including buttons and pieces of uniforms—sometimes at inopportune times, such as on formal parades. Rebellious Bobbies occasionally butted unsuspecting soldiers.

The Royal Warwickshire Regiment became the Royal Regiment of Fusiliers in 1968. The last Bobby passed away in 2005. However, the image of Bobby the Antelope can still be seen on Fusiliers' dress uniforms.

Montgomery was 21 when he began his career with the Warwicks. According to mascot handler Ian Spooner, Montgomery said during a 1968 conversation that he started his career as one of Bobby's handlers. Montgomery had the regimental symbol of Bobby the Antelope tattooed on his left arm.

He later became the Colonel of the regiment, an honorary position bestowed by the British sovereign.

A "Sword and Shield" Warrior: A Unique Combination

Montgomery possessed a combination of intellectual power and raw fighting spirit seldom seen in history. Montgomery was both an aggressive, hands-on warrior and a skilled administrator—a truly rare combination.

President Franklin Roosevelt conferred to Montgomery the highest honor of Chief Commander under the Legion of Merit, a U.S. military award. Although the Chief Commander is normally reserved for heads of state, Roosevelt used his authority in an Executive Order to present this honor to some Allied World War II leaders who "distinguished themselves by exceptionally meritorious conduct in the performance of outstanding services." The 13 stars are arranged in a pattern that is on the Great Seal of the United States. Image courtesy of Wikimedia Commons.

Polish president in exile Wladyslaw Raczkiewicz (middle left) during a military ceremony in World War II. The Polish government was based in London at the time. Photo courtesy of Wikimedia Commons.

Although World War II was still being fought, Montgomery was honored in October 1944 by President Raczkiewicz who conferred upon him the decoration of the Virtuti Militari V (Silver Cross) Class, given to military commanders who exercised daring and bravery. Notice above from *The London Gazette* on Oct. 27, 1944.

DECORATION CONFERRED BY THE PRESIDENT OF THE REPUBLIC OF POLAND.

Virtuti Militari V Class.

Field-Marshal Sir Bernard L. Montgomery, K.C.B., D.S.O. (8742).

The Russians also honored Montgomery for his wartime service. In June 1945, Soviet Marshal Georgy Zhukov decorated Montgomery with the Russian Order of Victory. The photo (courtesy of Wikimedia Commons) above shows Montgomery (left) next to Dwight D. Eisenhower, and Zhukov (center) with others attending the ceremony in Frankfurt. The Order of Victory was a rare honor only awarded 20 times. Montgomery was one of only five foreigners ever to receive the order, reserved for generals and marshals who have conducted successful combat operations in at least one army group or several fronts that led to a radical change in favor of the Red Army. Montgomery is the only British military leader to ever receive this rare distinction, now displayed at the Imperial War Museum in London. The notice above appeared in *The London Gazette* on June 21, 1945. The Order of Victory is shown below. (Note: Although the *Gazette* lists the "Paratroop" Regiment, this regiment is normally referred to as the Parachute Regiment.) Photo courtesy of Wikimedia Commons.

DECORATIONS CONFERRED BY THE PRESIDIUM OF THE SUPREME COUNCIL OF THE UNION OF SOVIET SOCIALIST REPUBLICS.

Order of Victory.

Field-Marshal Sir Bernard Law MONTGOMERY, G.C.B., D.S.O. (8742), late The Royal Warwickshire Regiment, Colonel Commandant, Paratroop Regiment.

Throughout history, there are few examples of "fighting generals" who also have intellectual gifts for administration. For example, the renowned Chinese strategist Sun Tzu had a powerful strategic intellect but was not known for his battlefield exploits. By contrast, German Field Marshal Erwin Rommel excelled as a leader in combat, but had a limited appreciation for military theory and was frequently

Hannibal crossing the Alps on elephants, painting attributed to Nicolas Poussin (1594-1656), image courtesy of Wikimedia Commons.

at odds with his administration. Both men were very successful yet had opposite personalities.

This contrast is perhaps best exemplified by two Roman military commanders during the Second Punic War.

This conflict saw the Romans competing for control of Italy against the North African armies of Hannibal. During this war, two leaders became famous for their divergent personalities and successful approaches to battle. Quintus Fabius Maximus exemplified the personality of the scholarly strategist and administrator; he chose to avoid open combat and instead wore the enemy down through "delaying" tactics and logistical methods. By contrast, Marcus Claudius Marcellus was the epitome of the forceful battlefield general—winning spoils in single combat and sacking the city of Syracuse. Both men were efficient but fundamentally different. In recognition of their different charisma, Fabius was dubbed the "Shield of Rome," while Marcellus became popularly known as the "Sword of Rome."

In his biography of Marcellus, the ancient historian Plutarch described the difference in the two leaders' fighting style:

Statue of Fabius Maximus in Vienna, by Joseph Baptist Hagenauer (1732-1811), photo courtesy of Wikimedia Commons.

"Fabius Maximus...was held in the greatest esteem for his sagacity and trustworthiness, his excessive care in planning to avoid losses was censured as cowardly inactivity. The people thought they had in him a general who sufficed for the defensive, but was inadequate for the offensive, and therefore turned their eyes upon Marcellus; and mingling and uniting his boldness and activity with the caution and forethought of Fabius, they sometimes elected both to be consuls together, and some made them, by turns, consul and proconsul, and sent them into the field. Poseidonius says that Fabius was called a shield, and Marcellus a sword. And Hannibal himself

used to say that he feared Fabius as a tutor, but Marcellus as an adversary; for by the one he was prevented from doing any harm, while by the other he was actually harmed."

Rarely in history have those two different fighting spirits converged in one man. Montgomery was one such man.

Vigorously athletic and competitive, Montgomery prized physical toughness and excelled as an infantryman. He was Battalion Sports Officer and also won prizes in man-to-man bayonet fighting contests in the gymnasium before World War I. He was also a very enthusiastic cross-country skier. He continued to participate in and excel at athletic activities throughout his life.

However, witnessing the disastrous military failures of World War I led him to become involved in military administration. He sought opportunities to enhance his education. His interests in military history, theory, literature, and human psychology helped him realize his goals in action with brilliant success. Montgomery was able to deliver hard punches to battlefield enemies while managing his forces deftly.

In this respect, Montgomery shares much in common with the notable Japanese swordsman and military theorist, Miyamoto Musashi. Like Montgomery, Musashi was a physically aggressive man who devoted his entire life to mastering the art of war. Also like Montgomery, Musashi was an independent thinker who believed a warrior needed to have a well-rounded intellectual background in order to truly master strategy. Both Montgomery and Musashi placed heavy emphasis on human psychology.

The First World War

Perhaps the greatest impact on Montgomery's life and military philosophy was his experience in World War I. During interviews and in his writings, Montgomery expressed that World War I changed his outlook on life. He became driven to master war as a professional and had ambitions to lead military reforms.

Soldiers of Montgomery's Royal Warwickshire Regiment rest in exhaustion in July 1916 with their rifles stacked on a grassy field during the Battle of the Somme, July 1 to Nov. 18, 1916. Montgomery took part in this battle with his regiment. Photo courtesy of Wikimedia Commons.

Montgomery, aged 26, arrived on the frontlines in 1914. He served actively throughout the entire war on the Western Front except for the time he spent recovering in hospital after being wounded. Montgomery immediately distinguished himself for his courage during his first major battle at Meteren on Oct. 13, 1914—the young soldier received the Distinguished Service Order for "conspicuous gallant leading," engaging in fierce bayonet fighting and successfully routing Germans.

His gallantry nearly cost him his life. A German sniper shot him through the right lung while he was fighting near some houses. It should have been a fatal injury. Montgomery fell in the open. A member of his platoon, who evidently had some knowledge of combat medicine, rushed over and began to dress the wound. Montgomery's rescuer was shot through the head and killed by the sniper. Montgomery was pinned beneath the corpse, which received several more shots intended for him. Another shot hit Montgomery in the knee. His platoon presumed he was dead.

"Inside a World War I field hospital with gassed and wounded soldiers on stretchers," by Eric Kennington, who enlisted with the London Regiment and became a war artist with the Ministry of Information (1918). Image courtesy of Wikimedia Commons.

He was later picked up by stretcher-bearers and taken to an Advanced Dressing Station. Medical staff expected him to die; a grave was dug for him. However, Montgomery stubbornly clung to life and made a miraculous recovery. As he recuperated, he had time to reflect on his experiences.

> *"When I lay in hospital, pondering over this matter, I came to the conclusion that war was a highly professional business, and there is no room in war for the amateur,"* he later said, during a 1960s interview at Memorial University of Newfoundland with Lord Stephen Taylor of Harlow. *"So I decided that I must study my profession and get right down to it,"* he said. *"And I gave up everything—everything. I took no part in social life. I worked."*

Montgomery's decision to reject distractions and devote himself entirely to his profession is very similar to the approach of the swordsman Musashi, who chose life as recluse to perfect his skills. In both cases, their dedication proved effective; both Montgomery and Musashi made unparalleled achievements in battle.

In 1914, Montgomery began the Great War leading a platoon of 30 men. By the time the war ended in 1918, he had risen rapidly in command due to his skills and was the Chief of Staff of a Division.

"That was my first introduction to leadership of men in battle," he said.

The approaches Montgomery adopted as a commander were directly influenced by what he experienced during World War I. One example is that Montgomery's art of war is based on his belief in personal and encouraging contact with one's men. This is something he was denied as a young soldier.

A view of an abandoned British trench (circa 1914 to 1918). Montgomery was appalled by the horrific conditions on the frontlines of World War I and also by higher commanders' lack of interest in the experiences of their men. Photo courtesy of the Library of Congress.

Montgomery was very upset by the fact that generals and higher commanders appeared to take no interest in the circumstances of men on the frontlines.

> *"A remarkable, and disgraceful, fact is that a high proportion of the most senior officers were ignorant of the conditions in which the soldiers were fighting,"* he wrote of World War I. He also wrote, *"There was little contact between the generals and the soldiers."* *"The soldier can feel intense loneliness during moments in battle. In the early stages of the 1914/18 war, as a young platoon commander on a patrol at night in no-man's-land, I was several times cut off from my men. I was alone in the neighborhood of the enemy, and I was frightened; it was my first experience of war. I got used to it, of course. But in those days I came to realize the importance of the soldier knowing that behind him were commanders, in their several grades, who cared for him."*

Wounded British soldiers in a trench in Flanders, Belgium in 1918. Montgomery was nearly killed in 1914 at Meteren in northern France, near the Flanders region. After the war, he often quoted phrases from John McCrae's 1915 wartime poem, *"In Flanders Fields."* Photo courtesy of the Library of Congress.

As he rose in command, Montgomery ensured that he adopted a direct and hands-on approach to managing situations in the field.

On fighting in the trenches, Montgomery observed:

- *"Trench warfare was in fact siege fighting rather than open battle."*
- *"A rifleman in a stationary position could fire 15 rounds per minute accurately across the normal no-man's-land. The machine gun fired a continuous stream of bullets... Barbed-wire entanglements made another obstacle to approach...."*
- *"This was almost the only answer of the generals to the great power of the defensive given by machine guns, barbed wire, entrenchments and artillery—to attack with infantry in close formations in a direct charge across no-man's land, each soldier carrying almost half his own body weight.*
- *"Of all this, I was a witness: I suffered from it. I saw clearly that such tactics could not be the key to victory."*
- *"The Germans introduced gas of various types: asphyxiating, lachrymatory and vesicant (blistering). Mustard gas...was the worst, because it was the most unpleasant and disabling and took a long time to clear."*

Undoubtedly, Montgomery also formed his principles of morale based on what he experienced in the trenches.

"Fatigue and boredom in these ghastly conditions were as demoralizing as filth and danger," Montgomery recalled.

Montgomery was also extremely conscious of preserving the lives of his men on the battlefield and not gambling lives through brash or unnecessary risks. This also owed to his experiences in World War I.

The Royal Army Medical Corps aids wounded British soldiers of the 9th Division near Meteren in 1918. The sufferings that Montgomery witnessed during World War I and his near-death experience at Meteren changed his outlook on his life and career. Photo courtesy of the Library of Congress.

- *"The frightful casualties appalled me. The so-called 'good fighting Generals' of the war appeared to me to be those who had a complete disregard for human life,"* he wrote.

- *"It was normal for orders to be given that attacks were to be delivered 'regardless of loss'—often for several days in succession,"* he wrote.

- *"I was so horrified by what went on at Passchendaele and all these dreadful things,"* he said during an interview.

- Regarding the Battle of the Somme in which he participated in July 1916, Montgomery said that *"there were 30,000 men killed before lunch"* on one day.

He was evidently disturbed by the fact that many corpses were left unburied.

- *"A large number of those killed had no known grave, being merely blown to pieces by artillery fire; in some cases corpses formed part of trenches, being finally devoured by rats,"* he wrote.

As he rose in command, Montgomery subsequently ensured the humane burials of the dead, including enemy dead. Montgomery lost many comrades on the frontlines.

Dead Scottish soldiers lie in Flanders fields (circa 1914 to 1918). During World War I, decomposing bodies of soldiers on both sides remained scattered across battlefields and in trenches for years. Montgomery witnessed the remains of many men lying unburied. Photo courtesy of the Library of Congress.

- *"I myself fought side by side with splendid young men who did not know what they were in for but who offered their lives because we were all told by the political leaders that it was to be 'a war to end war.'"* He described the war as *"a dismal canvas, with very few bright spots"* and *"a tragedy."*

He was evidently deeply moved by John McCrae's famous poem *"In Flanders Fields,"* which he often quoted in his books. He most often quoted the last stanza, which is:

"To you from failing hands we throw

The torch; be yours to hold it high.

If ye break faith with us who die

We shall not sleep, though poppies grow

In Flanders fields."

This last part of the poem focuses on the duty of survivors to carry on the work of the deceased and not to "break faith" with them.

Montgomery's focus on these verses suggests that he perceived that he had a duty to make some kind of a difference based on the suffering he and his friends experienced.

Montgomery (circa 1940) as commander of the 3rd Division of the British Expeditionary Force. During World War II, Montgomery fought in many of the same regions in France where he first experienced war as a young man. Photo courtesy of Wikimedia Commons.

Clarity and Self-Control

Montgomery watches tank movements in North Africa in 1942. His personal tank driver in North Africa, Jim Fraser, related before his death in 2013 that Montgomery often told Fraser to stop the tank while driving on his rounds in the desert so Montgomery could talk to soldiers. They were surprised that an officer of his high rank would make time to converse with them. Photo courtesy of Wikimedia Commons.

As he devoted himself to pursuing perfection in the art of war, Montgomery developed a detailed philosophy of self-control. He expressed the view that a master of the battlefield required an "ice-clear brain" at all times.

Montgomery refused to smoke or drink alcohol during his life. This has resulted in his mistakenly being perceived as a religious puritan or teetotaler. These views are incorrect. Instead, Montgomery's views were part of his lifelong quest to master the profession of war, and—like Musashi—were doctrines he developed over time.

Montgomery, in fact, was punished by his parents for smoking at a young age, according to an interview he gave the BBC. As a young man, he refused his mother's demand to sign an anti-alcohol pledge. While stationed in India as a young officer for several years, Montgomery was socially obligated to engage in excessive drinking—which made him worry that he was losing his fighting edge, he wrote. His early experiences with the effects of alcohol led him to firmly reject it.

Montgomery was unable to smoke due to the after-effects of a gunshot wound to his lung from World War I. However, he was aware of the beneficial effects of cigarettes on soldiers on the frontlines. Many battle-weary men smoked to relax their nerves. He frequently distributed cigarettes to his men during World War II.

He rejected intoxicating substances for the sake of intellectual clarity. In his view, a good military leader should never compromise his ability to mentally focus.

"A Voluntary Aid Detachment nurse lighting a cigarette for a patient inside an ambulance during World War I," by Olive Mudie-Cooke (circa 1916). Image courtesy of Wikimedia Commons.

Religious Independence

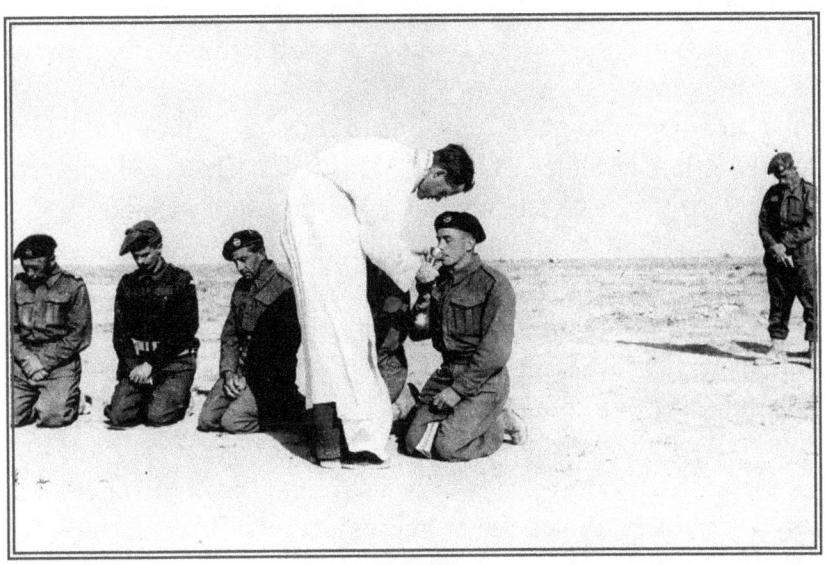

British soldiers participate in a religious service and receive Holy Communion from a priest in North Africa during World War II. Montgomery ensured that his men always had access to religious services and counsel. He believed spirituality was good for their wellbeing. Although he rejected hardline Christianity as practiced by his parents, Montgomery spoke and wrote of his private faith in God that seemingly deepened over time. Photo courtesy of the Library of Congress.

Montgomery had faith in God but was not dogmatically religious. This was in stark contrast to both his parents, whose entire lives revolved around church affairs. His father was a bishop.

"*He* [Montgomery's father] *took no part in the handling of the family. He was always communing with the angels—or whatever they do, bishops, you see,*" Montgomery said, with irreverent humor during a 1960s interview.

Instead of dedicating herself to her family, his mother preoccupied herself with the bishop's duties and beat her children violently with a heavy cane when displeased with them.

"*Certainly I can say that my own childhood was unhappy,*" Montgomery wrote. He added, "*My upbringing as a child had taught me to have resource within myself.*"

He was evidently unimpressed by the practice of religion he witnessed growing up. *"The need for a religious background had not yet begun to become apparent to me,"* he wrote of his state of mind as a young man.

Later in life, Montgomery came to view spirituality as a source of strength for soldiers. During World War II, he encouraged his men to pray and attend church services, and focused on the attributes of God as "all mighty" and "mighty in battle" to reassure them.

Always leading by example, he chose to participate in services and made prayer present in the soldiers' environments. He viewed religion as helpful to morale—which he referred to in writing as "the spirit of the warrior."

Although he expressed clear beliefs in God and found personal inspiration in the Bible, Montgomery's writings and speeches show that he rejected extreme interpretations of religion and practiced his own Christianity independently.

Conquering the Self

Montgomery believed that a great fighter should overpower his own thoughts and feelings to achieve high goals. In this, he had the same core approach expressed by Musashi: "By training you will be able to freely control your own body, conquer men with your body, and with sufficient training you will be able to beat 10 men with your spirit. When you have reached this point, will it not mean that you are invincible?"

> *"If you are going to handle men and women and control them in life, you've got first to learn to command and control yourself—in other words, to put it differently, to conquer yourself,"* Montgomery said. *"If you can't command and control yourself, conquer yourself, you won't be able to do this to other people. That's the first thing I learned."*

A tired-looking Bernard Montgomery briefly lets his guard down during a long day of smiling for the cameras in Denmark, 1945. Photo courtesy of the Museum of Danish Resistance Photo Archives.

One of Montgomery's key principles was projecting confidence during moments of personal uncertainty. This theme appears often in his writings. It implies that Montgomery privately experienced difficult moments during wartime, yet felt duty-bound not to show it—in fact, in accordance with the psychological warfare principles of Sun Tzu, he took the opposite approach. He considered it a commander's duty to overcome himself and "radiate" confidence to inspire others.

Publicity photos taken during World War II show that Montgomery put his theories into practice. While notable commanders on all sides of World War II cultivated martial frowns and scowls, Montgomery beamed cheerful smiles. Aware of the hardships being suffered by soldiers and civilians in Britain and Commonwealth countries during the war, Montgomery used his public persona to encourage others. Montgomery expressed the view that human emotions must *"be given an outlet in a way which is positive and constructive."*

Montgomery smiles during a publicity event in 1944. Unlike other generals, Montgomery did not cultivate a warlike public image. He appeared cheerful and confident to raise morale among soldiers and the general public, believing in a "positive and constructive" leadership style. Public domain photo.

One of the most impressive instances proving Montgomery's extraordinary self-mastery was his behavior immediately after his Miles Messenger aircraft crashed in August 1945. The small plane, carrying Montgomery and members of his staff, spiraled out of control while descending after the engine stopped running in midair. The rough crash landing left Montgomery badly injured. Battered and bruised, he sustained two broken lumbar vertebrae (a fractured spine). The fractures were severe enough to cause Montgomery back problems that lasted years afterwards. Yet immediately after this traumatic injury, Montgomery continued with his planned visit to troops of the 3rd Canadian Division.

Montgomery's downed Miles Messenger aircraft after a crash landing on Aug. 22, 1945. While other passengers were unharmed, Montgomery was left with several trauma injuries including spinal fractures. Images courtesy of the U.S. National Archives and Records Administration.

Shown here immediately after fracturing his spine in a plane crash, Montgomery calmly reviews troops and decorates soldiers. He made an effort to address the troops but had to cut the speech short because the pain was too great. His bravery and self-control were exceptional.

Despite the severe pain he must have felt, Montgomery inspected and decorated soldiers. He managed to maintain a calm demeanor, but was forced to cut his visit short as his pain became too great to manage. It is a testament to Montgomery's great self-control and willpower that he initially dismissed his personal injuries to proceed with his duties.

Despite his icy self-control, Montgomery was not emotionless nor was he detached from people around him. Some noteworthy military commanders in history have been extremely aloof; for example, the writings of T.E. Lawrence in *The Seven Pillars of Wisdom,* reflect a dispassionate and arguably callous attitude towards both British and Arab soldiers which suggests egocentrism on the part of Lawrence. This was not the case with Montgomery.

In spite of his philosophy of total self-discipline, Montgomery was not indifferent to others. For example, Montgomery notably suffered a minor emotional collapse in April 1945 in the wake of terrible Nazi atrocities he witnessed. On April 15, Montgomery and his men liberated the Bergen-Belsen concentration camp—the scene of some of the most gruesome Nazi torture in history. Most of the victims were women. Montgomery, who entered the camp with his soldiers, personally witnessed the appalling carnage. The horrifying scenes were enough to leave many British soldiers traumatized for decades, and surely affected Montgomery, who took special measures to provide immediate care to victims. Just a few days later, a loyal aide who served with him in North Africa and was recently engaged to be married was bayoneted by Nazi guerillas; the soldier was stabbed repeatedly and the Nazis left his gored remains to decay in a ditch. Witnessing these disturbing atrocities in such short succession affected Montgomery. He withdrew to his private quarters for two days and wept openly during a memorial service after emerging. However, he quickly recovered.

Montgomery personally cared about the welfare of the men serving under his authority. He believed in discipline but rejected cruelty. He took a warmhearted approach to commanding troops—bonding personally with his men as brothers-in-arms. Montgomery was known for his active acts of kindness to soldiers. In some instances, Montgomery wrote obituaries to honor his fallen comrades.

Montgomery gives reading material to a British soldier on a return trip from visiting troops near Kleve, Germany, 1945. Traveling in a DUKW vehicle, Montgomery distributed magazines, cigarettes and newspapers to his men at the front. Photo courtesy of the United States Holocaust Memorial Museum.

The same approach was also taken by the famed Korean Admiral Yi Sun Shin. Like Montgomery, Admiral Yi was highly successful in battle and immensely popular with his troops because of his personal kindness to his soldiers.

Montgomery wrote that one of his greatest heroes was Admiral Horatio Nelson.

> *"He knew how to win the hearts of men,"* he wrote of Nelson in 1961. *"He seemed to have a magnetic influence over all who served with him; he led by love and example."*

Mentorship

Montgomery jokes with men of the Royal Ulster Rifles during an inspection in early 1944. Montgomery, of North Irish and Scottish descent, had warm bonds of fellowship with Scottish and Irish soldiers. He wrote that one of the most memorable moments in his life was watching his Scottish troops march in a victory parade in Tripoli in 1943—*"every man an emperor,"* he wrote of their battle pride. He was on particularly good terms with men of the Ulster Rifles. "Being an Irishman himself he was very much at home with our soldiers and was impressed and affectionately amused by their humor, their independence and they way they got on with the job," according to a tribute published in 1978 the Royal Irish Rangers regimental magazine *"The Blackthorn"*. Photo courtesy of Wikimedia Commons.

In his writings, Montgomery expressed feelings of neglect growing up and lack of guidance in his early military career. He credited much of his success in life to teachers and older officers who were supportive during difficult times.

At times when his future path seemed uncertain, Montgomery claimed he was *"saved by good luck and good friends,"* and that he had *"required advice and encouragement from the right people to set me on the road, and once that was forthcoming it was plainer sailing."*

He wrote of his early difficulties in school.

> *"I had received no preparation for school life...in my case, once the intention and the urge was clear the masters did the rest and for this I shall always be grateful."*

Nearly expelled from Sandhurst, he was helped by *"one staunch friend among the Company officers, a major in the Royal Scots Fusiliers... He was my friend and adviser and it is probably due to his protection and advice that I remained at Sandhurst, turned over a new leaf and survived to make good."*

Consequently, Montgomery demonstrated a tendency to show compassion toward younger individuals—particularly troublemakers—and a willingness to provide fatherly encouragement and guidance.

"He could forgive a young officer or lance corporal most things as long as their misdemeanors were not dishonest or in bad taste. If you were over the age of 25 you were expected to know better and could expect little mercy," according to a 1978 tribute by Maj. Gen. H.E.N. "Bala" Bredin of the Royal Ulster Rifles.

Mentorship was important to Montgomery. He was always willing to give career advice and was often sought after for guidance. He regarded teaching as important and shared life lessons and military doctrines with others.

> *"I once said to an A.D.C. of mine who was ambitious, but inclined to be idle: 'Remember that without great toil those who triumph are very few in any profession,'"* he wrote.

"The Human Factor": Montgomery's Mastery of Psychology

Many of Montgomery's actions were premeditated and deliberate. He placed a great emphasis on human psychology to a degree not often seen among Western leaders.

Montgomery inspects Polish troops in the U.K., circa 1944. Photo courtesy of the Jozef Pilsudski Institute of America.

Montgomery often wore elements of different uniforms in order to relate to certain troops. His most notable wardrobe change was a black beret—an element of a tank crewman uniform. This gave him a very raffish look, never before seen among British commanders—most of whom tended to be solemn and wore stiff caps. In addition, Montgomery also proudly wore a tank crewman's badge on the cap in addition to his own commander's badge, which was unheard of.

Army authorities initially protested against Montgomery's unorthodox cap—however, it effectively popularized him with the soldiers and he was allowed to continue wearing it. The beret became so much part of Montgomery's public image that he was rarely seen without it during the war. After the war, Montgomery wrote that he adopted the beret for strategic reasons to influence his troops.

"To obey an impersonal figure was not enough. They must know who I was. This analysis may sound rather cold-

blooded, a decision made in the study. And so, in origin, it was: and I submit, rightly so," Montgomery wrote.

Montgomery led troops of different nationalities from countries, such as Poland, operating in exile in London during World War II. Polish paratroopers (above) training in Britain (circa 1942). Photo courtesy of Wikimedia Commons.

Montgomery applied the same logic to other forces he commanded. When commanding parachutists, he adopted paratrooper gear—including a paratrooper beret, cap badge, and silk jump scarf.

When commanding bomber crews, Montgomery adopted a leather bomber jacket. After the war, Montgomery used this tactic on a postwar diplomatic trip to Russia—he asked to be given elements of Russian clothing to wear during the visit. He ensured that photographs were taken in these situations.

This was an effective strategy, which also was sincere on Montgomery's part. His writings express strong convictions about the need to relate to others on a human level—a practice he emphasized in his command style.

Montgomery publicized images of himself being affectionate with his two puppies called "Hitler" and "Rommel" in early July 1944. Photo courtesy of Wikimedia Commons.

He also used psychology to disparage and demoralize wartime enemies. During World War II, he publicized his adoption of two puppies—one named "Hitler" and the other "Rommel." This was popular with the media and taken for a joke. However, this was surely deliberate. At a time when Adolf Hitler and Erwin Rommel were among the most feared men in the world, Montgomery made a show of associating their names with two small, fluffy dogs. This was most likely a tactic to belittle the enemy in the public eye.

Montgomery used another stratagem to undermine the Germans while fighting in North Africa—he invited one of Rommel's captured subordinates (General Wilhelm Ritter von

Montgomery (right) posed for a photo with General Wilhelm Josef von Thoma in 1942. Photo courtesy of Wikimedia Commons.

Thoma) to dinner and publicized the event to the media. Some took this at face value as a gesture of "gentlemanly" hospitality. However, the move was clearly methodical and designed to upset the Germans who were still waging a desperate war. Montgomery wanted to be seen having a "friendly chat" with a German general—which would hint at an intelligence leak, and likely upset Rommel and other enemy commanders. Further evidence of Montgomery's subterfuge is the fact that he made a recording of the "private" dinner conversation.

Montgomery used his talent at psychological ruses to crack German resolve and force their surrender on Lüneburg Heath in May 1945. When a German delegation arrived at his headquarters to discuss possible ceasefire terms, Montgomery cornered his wartime enemies into an unconditional surrender using a series of ploys. He forced the Germans to wait a long time for his arrival, stand under a British flag, and created the illusion of luxury in the rooms where meetings took place. He played upon the enemy's fears by gesturing with a map when speaking to them. He also arranged for Allied bomber planes to swoop menacingly overhead.

Montgomery (right) signs the Instrument of Surrender of the German Armies in northern Germany in May 1945. Public domain photo.

Although the Germans initially approached Montgomery intent on negotiation, they became completely demoralized and quickly gave up their fight. Their surrender resulted in the liberation of Denmark, Holland and many other regions from Nazi oppression.

Montgomery made a habit of studying his opponents' faces during battles as he formulated plans.

> *"I always had in my caravans during Hitler's war a picture or photograph of my opponent…I would study his face and see if I could fathom his likely reaction to any action I might set in motion; in some curious way this helped me,"* he wrote.

This was yet another very psychological approach to war.

The following Chapters 1 through 10 contain Montgomery's thoughts and writings on his approach to war.

—Zita Steele

CHAPTER 1:
CORE PRINCIPLES

Portrait of Bernard Montgomery taken during World War II. Public domain image.

- *My reading over the years has convinced me that nobody can become a great commander, a supreme practitioner of the art of war, unless he has first studied and pondered its science.*

- *Both study and practice are necessary: first, a study of the science of war, and secondly, learning to apply the study*

Montgomery stops his car to talk with troops under his command near Caen on July 11, 1944. Photo courtesy of Wikimedia Commons.

practically in battle.

• *War is not an act of God. War grows directly out of the things which individuals do or fail to do. It is, in fact, the consequence of national policies or lack of policies.*

• *Success does not happen; it is planned.*

• *A man's ordinary day-to-day life must be well-organized.*

• *Hardship and privation are the school of the good soldier; idleness and luxury are his enemies.*

• *Man is still the first weapon of war. In war, it is the man that counts, and not only the machine.*

- *In war, it is "the man" that matters.*

- *A good tank is useless unless the team inside it is well trained, and the men in that team have stout hearts and enthusiasm for the fight, so it is in all other cases.*

- *The great art is to be able to grasp rapidly the essentials of a military problem, to do something about it quickly, and to see that other people also do a good deal about it very quickly.*

- *It is most dangerous to seek for a code of tactics which will meet "the normal battle." There is no such thing as a "normal battle." Every battle is different.*

- *In any military organization there is no surer way to disaster than to take what has been done for many years, and to go on doing it—the problem having changed.*

- *Success is vital; but battles must be won with the least possible loss of life.*

- *Military problems are in essence simple; but the ability to simplify, and to select from the mass of detail those things and only those things which are important is not always so easy.*

- *A general must understand the mind of his opponent, or at least try to do so.*

- *Generals are meant to win battles, and the good general of today will do so with the least possible loss of life.*

- *A good military leader must dominate the events which encompass him. Once events get the better of him he will lose the confidence of his men, and what that happens he ceases to be of value as a leader.*

Montgomery stops his car in 1943 to offer cigarettes to troops in Italy. Photo courtesy of Wikimedia Commons.

- *All soldiers will follow a successful general.*
- *With good men, anything is possible.*
- *Nothing will be accomplished in the crisis by a man without a sense of duty.*
- *Men require to be united if they are to give of their best.*
- *The troops must be brought to a state of wild enthusiasm. They must enter the fight with the light of battle in their eyes and definitely wanting to kill the enemy. The spoken word from the commander to his troops is far more effective than any written matter.*

Scottish Highlanders of the British 8th Army in Tripoli discover an ancient Roman bath at Leptis Magna, Tripolitania, and begin to put it to use during a rest period in the North Africa campaign. Photo courtesy of the Library of Congress.

- *During months of hard fighting, I have found that no two problems are ever the same.*

- *A commander must keep an open mind, consider the conditions of the problem very carefully, and decide on a method suitable to the occasion.*

- *Control in battle is largely dependent upon good communications. It is absolutely essential that these should be 100% efficient.*

- *An obstacle loses 50% of its value once the enemy can reconnoiter it.*

- *A sure way to victory is to concentrate great force at the selected place at the right time and to smash the enemy. Dispersion of effort—and of resources—is fatal to success.*

- *Comradeship and teamwork are of enormous value in all units. If a man feels he is a stranger, his morale and fighting value will be low.*

- *If every man can be made to feel that he is a member of one happy and successful family, and that the honor and good name of his unit and division are worthy of any sacrifice which he may be asked to make; a commander can be sure that his division will respond to every call made upon it.*

CHAPTER 2:
THE WAY OF THE COMMANDER

Montgomery supervising a training exercise in preparation for the Allied Rhine crossing in 1945. Photo courtesy of Wikimedia Commons.

- *When all is said and done, the crucible of war will determine the fine metal of which a general is made.*

NEED FOR CLEAR THOUGHT:

- *A commander has got to be a very clear thinker. He is likely to fail unless he has an ice-clear brain at all times and a disciplined mind; this implies being abstemious, particularly in things such as drinking or smoking.*

Nearby soldiers closely observe Montgomery as he participates in victory celebrations in Denmark, 1945. Photo courtesy of the Museum of Danish Resistance Photo Archives.

- *The leader must continue to think longer than his men, and his thoughts must lead to action.*

- *A commander must spend a great deal of time in quiet thought and reflection, in thinking out the major problem, in thinking how he will defeat his enemy.*

- *The wise commander will see very few papers or letters. He will refuse to sit up late at night conducting the business of his army. He will be well advised to withdraw to his tent or caravan after dinner at night and have time for quiet thought and reflection. It is vital that he keep mentally fresh.*

- *No officer whose daily life is spent in considering details, or who has not time for quiet thought and reflection, can make a sound plan of battle on a high level or conduct large-scale operations efficiently.*

Leadership:

- *The best type of leader earns the respectful admiration of his men because he possesses certain good qualities which they lack.*
- *A brutal leader who disregards the feelings of his men will not infuse them with the quality of self-respect; the morale of the troops he commands, regardless of his qualities as a leader, will not be of the highest.*
- *A leader should be efficient. He should possess self-confidence. He should be firm and just in his dealings with his men. He should be clear-cut and definite in giving his orders. He should pay attention to administrative details. He should prepare his men in advance for any new experience they may have to meet.*
- *If the leader will decide, the men will follow and will fight. If there is indecision, they will hesitate and will flee. In short: "fight and survive"; "fear and be slain"; the leader decides.*
- *Men recognize in their leader some quality which they themselves do not possess; that quality is decision.*
- *A leader's character will develop in proportion to the responsibility with which he has been entrusted.*

Montgomery, standing on a jeep, addresses troops in England in February 1944. Montgomery made thorough efforts to ensure that all troops participating in the D-Day invasion had personal contact with him before the attack began. Photo courtesy of Wikimedia Commons.

• *A general must never be reluctant in allotting praise where it is due.*

• *The general who looks after his men and cares for their lives, and wins battles with the minimum loss of life, will have their confidence.*

For Commanders:

• *A commander must constantly study the state of morale so that he may seize the right psychological moment to initiate a series of talks to his troops.*

• *A commander must gain the complete trust and confidence of his men. There is no book of rules which will help him in this matter. Each commander will adopt his own methods* and the ones best suited to his own makeup. Success in battle will produce quick results.

• He must keep his finger on the spiritual pulse of his armies.

• Commanders must constantly talk to the troops under their command. The more that troops can see their commanders, hear them talk, and be put "in the picture" by them, the higher will be their morale.

• A commander must understand that bottled up in men are great emotional forces which have to be given an outlet in a ways which are positive and constructive, and which warms the heart and excites the imagination.

• I do not believe that a commander can inspire great armies, or single units, or individual men and lead them to great victories unless he has a proper sense of religious truth.

• A commander must watch carefully his own morale. The battle is in effect a contest between two wills—his own and that of the enemy commander. If his heart begins to fail him when the issue hangs in the balance, then the enemy commander will probably win.

• Commanders in all grades must have qualities of leadership. They must have the initiative, and they must have the "drive" to get things done. They must have that character and ability which will inspire confidence in their subordinates.

• Commanders must above all have that moral courage, that resolution, and that determination which will enable them to stand firm when the issue hangs in the balance.

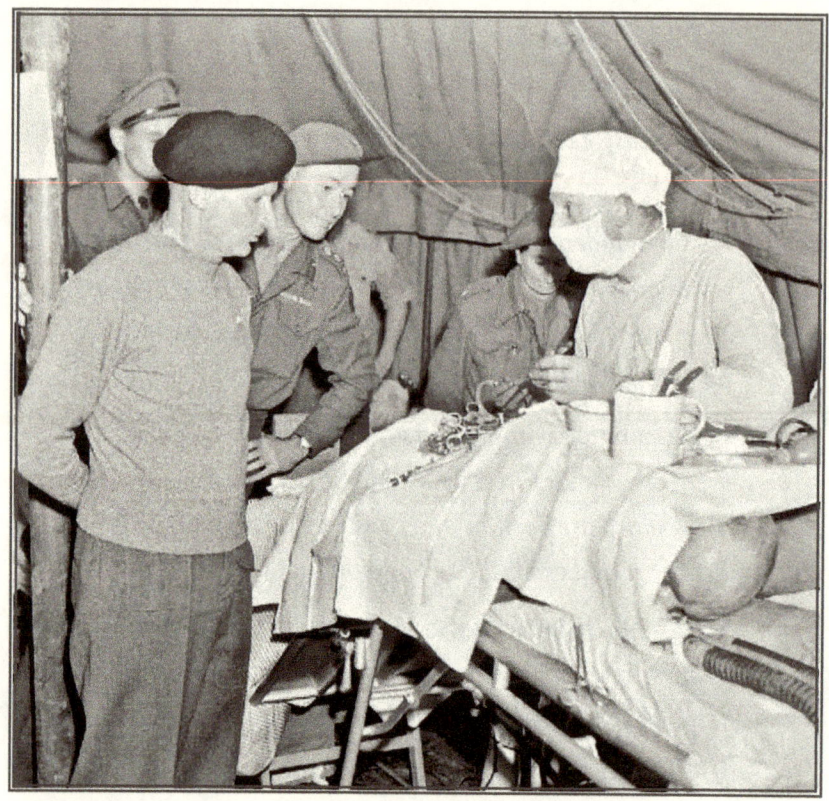

Montgomery observes a surgeon treating a soldier's bullet wound in France in June 1944, not long after the Normandy landings. Montgomery was accustomed to visiting wounded soldiers in hospitals. As well as taking a sincere interest in his troops' wellbeing, Montgomery also believed a commander needed to present a calm and reassuring demeanor. Photo courtesy of Wikimedia Commons.

> • *Probably one of the greatest assets a commander can have is the ability to radiate confidence in the plan and operations, when inwardly he is not too sure about the outcome.*

Importance of Calmness:

- *The two vital attributes of a leader are decision in action and calmness in crisis. With them, he will succeed, and without them he will fail.*

- *The leader's power of decision results from his ability to remain imperturbable in the crisis. His calmness prevents panic, and his resolution compels action.*

- *A leader must be less fearful than his men. He need not be impervious to fear, since men require a human figure to lead them. What he must do is to radiate an atmosphere of confidence which will show his men that he is less afraid than they. He must have the moral courage to stand firm when his men are wavering. In this respect they will judge him by his power of thought and action in a crisis.*

- *The leader's greatest asset is the ability to act normally in abnormal conditions, to continue to think rationally when his men have ceased to think, to be decisive in action when they are paralyzed by fear.*

- *He has got to inspire confidence in his soldiers on the battlefront, and in the general public in the home country.*

- *A commander has a unique opportunity to influence personally the morale of his division. The surest way to get a high morale is to instill confidence.*

- *If the troops have complete confidence in their commander, then all is well, since they know that he will see to everything.*

Montgomery addressing men of the Polish 1st Armored Division in Scotland (circa 1944). Photo courtesy of the Jozef Pilsudski Institute of America.

On Handling Troops

- *One of the first responsibilities of a commander in the field is to create what I would call "atmosphere", and in that atmosphere his staff, his subordinate commanders and his troops will live, and work, and fight.*

- *The commander should ask himself what are the things that really matter, that form the basis of the whole operation, that must be got right.*

- *He should be known to as many as possible of all ranks serving under him. This can best be achieved by short talks to all ranks during periods out of action on subjects which he wishes to put across himself, by frequent visits to brigades and units both in and out of action, and by personal messages on special occasions.*

- *The lives of his men should be precious to any commander in battle; they are not to be risked without cause, nor used when other means will serve.*

  —To carry out any operation, a commander has certain resources. These are:

  —The number of men available for the operation.

  —The physical ability of these men.

  —The mental ability of these men.

  —The material, arms, ammunition, vehicles, petrol, supplies of all kinds, with which these men will fight.

- *A commander has at his disposal certain human material; what he can do with that material will depend entirely on himself.*

Subordinates:

- *Probably one of the most important requirements in a commander is that he must be a good judge of men. He must be able to choose as his subordinates men of ability and character who will inspire confidence in others.*

- *A commander must be a good judge of men, and be able to have the right man in the right place at the right time.*

- *A commander must know in what way to give verbal orders to his subordinates. No two subordinates are the same; each will require different treatment. Some will react differently from others. All this must be known.*

- *Operational command in the field must be direct and personal, by means of visits to subordinate HQ, where orders are given verbally. It is quite unnecessary to confirm these orders in writing. Subordinates who cannot be trusted to act on clear and concise verbal orders are useless.*

Montgomery surrounded by his senior officers at Eighth Army headquarters in Italy during World War II. Montgomery believed that assembling an efficient team of subordinates was essential to military success. Photo courtesy of Wikimedia Commons.

- *Eventually a mutual confidence will grow up between the commander and his subordinates; once this has been achieved there will never be any more difficulties or misunderstandings.*

- *A commander must train his subordinate commanders, and his staff, to work and act on verbal orders and instructions.*

- *A commander must carefully watch his subordinate commanders for any signs of overstrain, or staleness, in order that he may rest or replace them before it is too late.*

Montgomery in a Stuart command tank in North Africa in March 1943. Photo courtesy of Wikimedia Commons.

In Battle:

- *Generally speaking, it may be said that there are two things a commander must do:*

    1. Create the fighting machine and forge the weapon to his liking.

    2. Create the Headquarters organization (or set-up) that will enable the weapon to be wielded properly and develop its full power rapidly.

Montgomery (top left) stands on a jeep with his back to the camera as he addresses the Polish 1st Armored Division in Scotland (circa 1944). Montgomery's reflections on high command derived from hard practical experience in the field. In 1944, he commanded an estimated 2 million men from different nations. Photo courtesy of the Jozef Pilsudski Institute of America.

- *It is absolutely vital that commander should keep himself from becoming immersed in details. If he gets involved in details, he will lose sight of the essentials that really matter: he will be led off on side issues that will have little influence on the battle, and he will fail to be that solid rock on which his staff must stand.*

- *A main responsibility of a general is to organize training, not only for the campaign which is to be opened up but also for any particular battle during the campaign.*

- *The divisional commander has got to keep a tight grip on the tactical battle, and he cannot do this unless he knows what is going on. But he must not wait for information which may never arrive. He must go forward and find out the situation for himself.*

- *By study of intelligence and by reconnaissance, the commander will then be in a position to make his plan which must be simple. He will decide to what extent to soften up the enemy by preliminary artillery and air bombardment, and whether to attack by night or day.*

- *The commander must foresee his battle: he must decide in his own mind, and before the battle starts, how he wants the opportunities to be developed. He must then use the military effort at his disposal to force the battle to swing the way he wants.*

- *There is far too much paper in circulation in the Army as a whole; no commander can have time to read all this paper and also do his job properly. Much of the paper in circulation is not read; much of it is not worth reading.*

- *His armies must know what he wants. They must know the basic fundamentals of his policy and must be given firm guidance and a clear "lead." Inspiration and guidance must come from above and must permeate throughout the force. The whole force will thus acquire balance and cohesion, and the results of the day of battle will be very apparent.*

- *The higher commander himself should stand right back and have time to think. His attention should be directed to ensuring that the basic foundations and cornerstones of the plan are not broken down by the mass of detail that will occupy the attention of the staff.*

Montgomery reviewing a plan with a staff officer in Tripoli in January 1943. With previous experience as both a staff officer and a frontline infantryman, Montgomery demonstrated considerable talent at leading battles and his administration. The combined ability to fight and administrate well is rare among history's great military commanders. Photo courtesy of Wikimedia Commons.

- *He must decentralize. He must lay down "the form" very clearly. He must then trust his subordinates, and his staff, and must leave them alone to get on with their own jobs.*

- *He himself must devote his attention to the larger issues. He must not "bellyache" about details.*

- *A higher commander cannot often speak to his troops personally. He can, and should, speak to officers collectively whenever suitable opportunities exist.*

- *Though he cannot often speak personally to his troops, he can keep in touch and get his personality across by means of personal messages. Before any big operation, and at other times such as Christmas, a personal and inspiring message from the C-in-C will be of great value. Such messages must be drafted very carefully; they must be exactly right. They must not be too frequent but should be kept for very special occasions.*

- *My final advice to any officer who may be called on to exercise high command in war is as follows:*

    1) Have a good Chief of Staff.

    2) Go for simplicity in everything.

    3) Cut out all paper and train your subordinates to work on verbal instructions and orders.

    4) Keep a firm grip on the basic fundamentals—the things that really matter.

    5) Avoid being involved in details, leave them to your staff.

    6) Study the factor of morale. It is the big thing in war and without a high morale you can achieve nothing.

    7) When the issue hangs in the balance radiate confidence in the plan and in the operations, even if inwardly you feel not too certain of the outcome.

    8) Never worry.

    9) Never bellyache.

    10) Keep fit and fresh, physically and mentally. You will never win battles if you become mentally tired, or get run down in health.

Montgomery receives a German delegation for surrender at Lüneburg Heath on May 3, 1945. Montgomery used a series of psychological stratagems to knock the Germans off balance and direct the negotiations the way he wanted (see Introduction). Photo courtesy of Wikimedia Commons.

The Enemy:

- *The commander must seek to dominate the enemy before he launches his attack.*

- *Foresight and anticipation by the commander will reduce the time necessary for the mounting of the attack.*

Montgomery, wearing an Australian slouch hat, surveys the battlefield from the top of a Crusader tank in North Africa in September 1942. Australian slouch hats worked well to shade the eyes, while the brims could be easily adjusted to allow great aim and flexibility when shooting rifles. Photo courtesy of Wikimedia Commons.

> - *The commander has got to strive to read the mind of his opponent, anticipate enemy reactions to his own moves, and take quick steps to prevent any enemy interference with his own plans; he must aim to be always "one move" ahead of his opponent. He has got to be a very clear thinker and able to sort out the essentials from the mass of factors that bear on every problem.*

- *The commander and his subordinates should be continually thinking out new "devilments" to harry the enemy.*

- *The exact method that a commander will adopt in order to set about his enemy will depend on varying circumstances.*

Remains of dead World War I soldier hanging on barbed wire in No Man's Land (circa 1914 to 1918). Photo courtesy the Library of Congress.

The Need to Bury the Dead:

- *Nothing rots morale quicker than the suspicion that a commander is careless of his men, which can grow from sensitive feelings as those aroused by indiscriminate burials of the dead and the sight of scattered graves neglected and bodies lying in ditches.*

- *No commander can afford to overlook the reverent and fitting burial of the dead—including enemy dead.*

- *Morale can be adversely affected by indiscriminate burials. Casual graves, scattered in ditches at the side of the road, must be avoided. The establishment of a cemetery and the proper and fitting burial of the dead are most important.*

- *The quality which maintains human dignity in battle and at the same time develops man's heroism is high morale.*

CHAPTER 3:
THE SPIRIT OF THE WARRIOR

Montgomery addressing frontline troops in Sicily, 1943. Introducing himself as "Monty," he would joke and talk sports with his men while giving them a breakdown of his larger strategy. Photo courtesy of Wikimedia Commons.

- *While the factors of command and control play a large part in the winning of battles, the greatest single factor making for success is the spirit of the warrior.*

- *Morale is the most important single factor in war. It is the quality without with no war can be won.*

British soldiers operate a heavy gun during World War I. Montgomery observed many soldiers tested by fear, panic and grim conditions during World War I, which formed the basis for many of his principles about morale. Photo courtesy of the Library of Congress.

What Is Morale?

- *Morale is a mental and moral quality. It is a quality peculiar to human beings because it is essentially the product of a mind with a conscience. It is that which in battle keeps men up on humanity's level. But humanity's level is not enough, because the strongest human instinct is the instinct for survival.*

- *Morale is also that which develops man's latent heroism so that he will overcome his desire to take the easy way out and surrender to fear.*

- *High morale is a quality which is good in itself and is latent in all men. It maintains human dignity. It enables fear and fatigue to be overcome. It is involved with the idea of conscience.*

Australian soldiers practice field exercises in October 1942. Montgomery had an Australian upbringing, spending his formative years growing up in Tasmania. He had great affection for his Australian troops. Photo courtesy of Wikimedia Commons.

- *A high morale is based on discipline, self-respect, and confidence of the soldier in his commanders and in his weapons.*

- *Without a high morale, no success can be achieved—however good may be the strategic or tactical plan, or anything else.*

- *The maintenance of the offensive spirit of a division is one of the primary tasks of its commander. He must ensure that they never settle down to a policy of "live and let live." In defense particularly, any tendency to be satisfied with an attitude of laissez-faire will inevitably lead to enemy domination of the area and invite subsequent disaster. To guard against this, the commander should be continually thinking how he can rest and relieve his troops and provide some change and relaxation for them. The more difficult the conditions, the more necessary this becomes.*

Royal Army Medical Corps soldiers carry an injured man to an ambulance after an enemy attack on an ambulance convoy in North Africa during World War II. Photo courtesy of the Library of Congress.

The Spiritual Test of Battle:

- *In battle, men who have kept a firm grip upon themselves will appear clean and vital in their appearance, while those who have gone to seed will be slovenly.*

- *In war, the moral stature of some men increases and their characters grow stronger and more closely-knit in proportion to the discomforts and dangers which they are called upon to face. Such men will occasionally perform in battle remarkable acts of selfless courage and daring, and will endure with extraordinary fortitude and patience the burdens thrust upon them.*

British soldiers recoil from shellfire in a trench in August 1916. Montgomery wrote that although fear is a normal and healthy reaction to danger, soldiers must learn to overcome it in order to succeed in battle. He based this on his personal experiences. Photo courtesy of the Library of Congress.

> • *Other men, however, will under the stress of hardships or dangers surrender to fear or fatigue and will allow their characters to disintegrate. This disintegration will usually take the form of a loosening of the moral fiber which results in timidity of action and slackness in appearance.*

Fear:

- *There are two aspects of fear. Fear can suddenly attack a man through his imagination. A corpse in a ditch or a grave by the side of the road will remind him of the peril of his position. He will suddenly realize that he himself is liable to be killed.*

- *Timid officers will be found during quiet periods in the line, groveling in the filth of some cellar while their signalers and runners, separated from the line, attempt to do their jobs by the flame of a guttering candle. In these latter cases there has been a general loosening of the character due to a partial surrender to fear.*

- *In extreme cases, men who have become afraid have sunk to the level of beasts. No longer in full control of themselves, they have become as sheep or rabbits, unable to act alone or think rationally. Their instincts have become those of the herd. They are either paralyzed by fear or gripped by unreasoning panic. Such men are exceptions, but they are a reminder of the value of high morale.*

- *Fear can also creep upon a man during periods of monotony in the line. At such a time he will have the opportunity to appreciate the dangers which beset his life. Fear acting through his thoughts can so reduce the man's hard core of courage that he will become nervous and fearful.*

- *The basis of fear is the awareness of danger. In itself this is healthy; for a man who is aware of danger automatically takes steps to provide against it. It is only when fear dominates the mind that it becomes unhealthy and leads to panic.*

American soldiers huddle together as they cross the Rhine under enemy fire in March 1945. Montgomery commanded American troops after the Normandy landings in 1944. Montgomery wrote that men in battle feel most afraid when they face danger alone—they derive fighting strength from unity and comradeship. Photo courtesy of Wikimedia Commons.

- *Man becomes aware of danger when he feels himself opposed to something more powerful than himself; in other words, when he feels that his own armament is unequal to that of the enemy who oppose him. The method by which the conquest of fear is achieved is the unifying of men into a group or unit under obedience to orders.*

- *The good soldier—the man with high morale—has not surrendered to fear and has maintained his personal standards. The bad soldier—the man with low morale—has become incapable of independent action and has to some extend shed a part of his human individuality.*

- *Fear makes men sluggish and indecisive, unable to decide of act for themselves.*

- *Fear destroys the faculty of thought and paralyzes action.*

- *A man alone is a man who will find it hard to stand up to the dangers of the line; a man alone is a man afraid.*

- *All men are afraid at one time or another and to a greater or lesser extent. In moments of fear they band together and look for guidance; they seek for a person to give decisions; they look for a leader.*

- *The leader's power over his men is based on his ability to cut through this "fear paralysis" and, in so doing, to enable others to escape from it.*

- *It is important to a man to lose his individual feelings and to become an integral part of the battalion, division and to which he belongs. The larger the unit of which he feels himself to be a member, the larger will be the estimation of his own armament and the less will be his fear.*

British infantry under Montgomery's command charge with bayonets during the Battle of El Alamein in 1942. Within a few months of taking command, Montgomery transformed men of the Eighth Army from a beaten and demoralized group into a fierce and highly competent fighting force. Photo courtesy of Wikimedia Commons.

Qualities Mistaken for Fighting Spirit:

- *Morale should not be confused with fitness or happiness or toughness.*

- *It is not a contentment or satisfaction bred from ease or comfort of living. Both of these contain a hint of complacency, and acceptance of luxury as an end in itself.*

- *High morale implies essentially the ability to triumph over discomforts and dangers and carry on with the job.*

British soldiers spar with French soldiers during a football match in 1940. Despite his own athleticism and love of sports, Montgomery did not view physical agility as the essence of good fighting spirit. Photo courtesy of the Library of Congress.

- *Nor is high morale achieved through fitness and healthiness alone. It is important not to confuse the idea of physical happiness with morale.*

- *The happy faces of men after a good game of football are not necessarily the faces of men with good morale.*

- *High morale is not happiness. Happiness may be a contributory factor in the maintenance of morale over a long period, but it is no more than that.*

- *A man can be unhappy but can still, regularly and without complaining, advance and defend within the terms of the definition.*

- *Morale is a mental rather than a physical quality, a determination to overcome obstacles, and instinct driving a man forward against his own desires.*

- *High morale is not toughness.*

A private in the British Army trains with a fixed bayonet and practices assault in Britain in 1944. Photo courtesy of Wikimedia Commons.

Toughness:

- *Toughness is a physical and not a mental asset.*
- *Some very tough men in war have turned out to be very disappointing in action.*
- *Tough men will occasionally perform an isolated act of bravery. Morale, however, is not a quality which produces a momentary act. It influences behavior at all times.*

The Four Pillars of Morale

- *It is now necessary to consider what factors constitute the morale of the soldier in the heat of battle. Certain factors may be described as essential condition without which high morale cannot exist. These four basic factors are:*

    1) Leadership,

    2) Discipline,

    3) Comradeship, and

    4) Self-respect.

- *A 5th factor—devotion to a cause—must exist but need not necessarily influence all the soldiers. Finally, there are numerous contributory factors which are of great importance but are not essential conditions.*

Riding atop a DUKW, General Montgomery salutes his smiling troops in the streets in Italy on Sept. 3, 1943. Photo courtesy of Wikimedia Commons.

1. Leadership in Morale

- *Morale is, in the first place, based on leadership.*
- *Men who are not properly led cannot be said to have good morale.*
- *Good morale is impossible without good leaders.*
- *This quality of leadership must be studied.*

A Gurkha soldier of the British Army during World War II. The fierce Gurkha warriors from Nepal were renowned for their fighting skills. Montgomery commanded Gurkhas in North Africa in addition to other soldiers of diverse nationalities. Photo courtesy of Wikimedia Commons.

- *Human beings are fundamentally alike, in that certain common characteristics apply to all men in varying degrees. In battle the most important of these characteristics is fear.*

- *The difficulties, dangers and discomforts inseparable from the battlefield make men cry out for the leadership they can do without in peace. At such moments men are too weak to stand alone; they find the burdens too great to bear and their own selves unequal to the task.*

- *The leader himself accepts the burdens of others and, by doing so, earns their gratitude and the right to lead them.*

Men of the Royal Tank Regiment share rations before going into combat. Photo courtesy of Wikimedia Commons.

- *Consider a platoon (i.e., 30 men) in action in the line. The men are drawn from all classes and of all types. They are there in the line because they have obeyed a long series of orders which it was easier to obey than to disobey. But now the test comes. It is easier for them not to obey orders; it is easier for them to stay where they are behind the hill and not advance over the crest into full view of the enemy who lies in wait beyond. The dominant motive force which drives them over the crest of the hill is their leader. It is his quality of leadership above all things which inspires the men to action.*

- *The leader's position as the man responsible for the lives and wellbeing of his men must be impressed upon him.*

- *In battle, the leader's preoccupation with his men's affairs will give him less time to think of his own fears.*

- *The mere fact of responsibility will increase the leader's powers of decision and make him confident of his ability to handle any crisis.*

- *Moreover, it is important to realize that while men are dependent on the word of a leader, they are capable of much independent action on their own and are even capable of taking their own decisions.*

- *What is required is that the leader shall give the initial and vital decision along which the men can work.*

- *If the leader gives the necessary orders, the men will carry them out magnificently. Any officer who has served in the line can produce many such instances. It is at these moments that the officer is amazed at the quality of the soldiers he leads.*

- *Good morale can be created in a narrow sphere by a good leader.*

- *Good morale implies confidence in the command and in the plan.*

- *A strong leader on the high level can have a powerful influence on the general attitude of the men of a platoon; he cannot, however, influence their movements over the last few vital yards of an attack. On the other hand, a strong leader on the low level can make his men carry out a single fine attack, but he cannot sustain their morale indefinitely, if there is a lack of confidence in the high command and its plan.*

Montgomery, in command of the 5th Corps, inspects men from the Suffolk Regiment on England's Channel coast in March 1941. Although Montgomery established a warm relationship with his troops, he maintained strong authority over them and instilled firm discipline, especially through rigorous training programs that he developed. Montgomery rose in the ranks after World War I due to his great skills at military analysis and at training soldiers for combat. Photo courtesy of Wikimedia Commons.

2. Discipline

- *A leader cannot do without discipline. His aim must be to create such a disciplined body of men that all his orders will be obeyed instantly.*

- *Soldiers must be treated with humanity and controlled by discipline. If the officer does this, he will gain the respect of his men at the same time he gives them self-respect.*

- *The object of discipline is the conquest of fear.*

- *Discipline strengthens the mind so that it becomes impervious to the corroding influence of fear.*

- *Discipline teaches men to confine their thoughts within certain definite limits. It instills the habit of self-control.*

- *It is a function of discipline to fortify the mind so that it becomes reconciled to unpleasant sights and accepts them as normal everyday occurrences.*

- *Men must be urged to fight fear with courage, so that they will advance or defend and not take refuge in flight or inaction.*

- *Discipline can help a man to lose his own identity and become part of a larger and stronger unit. It is in this way that discipline will conquer fear.*

- *Discipline seeks to instill into all ranks a sense of unity by compelling them to obey orders as one man. This obedience to orders is the indispensable condition of good discipline.*

- *Men learn to gain confidence and encouragement from doing the same thing as their fellows. They derive strength and satisfaction from their company. Their own identities become merged into the larger and stronger identity of their unit.*

- *Men must learn to obey orders when all their instincts cry out for them not to be obeyed. They must learn to obey orders in times of stress so that they will do so in times of danger. They must learn to carry out their tasks under any conditions and despite all difficulties. In this way the mass of loose individuals, with their fears and weaknesses, can be welded into a united whole, ready to act on the word of a leader.*

Montgomery and King George VI drive past cheering troops near Tripoli on June 21, 1943. Photo courtesy of Wikimedia Commons.

- *Discipline helps men to display fortitude in the face of fatigue and discomfort, while at the same time it helps them to conquer fear. It enables them uncomplainingly to triumph over difficulties which would have overcome them in times of peace. This constancy in enduring hardship and fatigue is the quality which is most frequently required of the soldier.*

- *Individual fortitude and corporate courage are the twin products of discipline.*

- *Discipline implies a conception of duty.*

A British Eighth Army officer stands near a disabled German 88 mill. gun in Cyrenaica during World War II (circa 1940 to 1943). When Montgomery took command of the Eighth Army, the men were totally demoralized due to constant defeats and were afraid of German Field Marshal Erwin Rommel. Montgomery waged a campaign to reenergize the army under his command and undercut Rommel's legend among the men. Photo courtesy of the Library of Congress.

- *For the soldier, this conception of duty does not embrace abstractions such as freedom or country or democracy. In battle, a soldier's sense of duty extends only to the friends who are around him. Abstractions are the sphere of the politician.*
- *This sense is instilled by discipline because it teaches men to obey orders as a matter of course, to know that is wrong not to obey them and right, that is their duty to do so.*
- *A certain type of training may induce men to go forward in attack simply out of fear of the consequences of not doing so. This applies only to the weakest and most feeble of men*

who are of little value in battle. This type of training is an essential part of discipline but it must never be mistaken for the whole.

• The type of training which implies a certain harshness and hardness has its value.

• Material comforts are now so insidious that there is some danger that this "old-fashioned" idea of discipline will be allowed to disappear. This must not happen.

• Soldiers will not win battles if their training has not been hard.

• The softening influence of civilian life must be replaced by the exacting demands of military training.

• Soldiers must forget the pleasures of peace and concentrate on the realities of war.

• Discipline seeks to conquer fear by welding men into a cohesive whole, united by obedience to orders.

• It aims to create a body strong enough to carry each of its members through dangers and difficulties which they themselves would be unable to face alone.

• The habit of obeying the leader's orders must be so instilled into his men that they will carry on and fight even though he himself falls. This aim cannot be achieved without discipline.

3. Comradeship

• War, though a hard business, is not necessarily a grim one. Men must laugh and joke together, must enjoy each other's company, and must get fun out of life even in times of danger.

Montgomery (right) laughs among his men during a visit from Prime Minister Winston Churchill in Caen in 1944. Photo courtesy of Wikimedia Commons.

- *Friendship causes men to give of their best.*

- *A man who has served among friends in the same platoon for a long time will be helped by them to face the trials of battle. He will feel all around him reserves of courage and purpose upon which he can draw. There will be a feeling of solidarity, and out of this feeling there will grow up a determination to advance together and defend together and even die together.*

- *A man must make friends in his platoon, friends whom he respects and admires. In battle these friends will prevent him from feeling lonely.*

A view of Montgomery's command tank in Italy 1944. Montgomery emblazoned his nickname "Monty" on the front of the tank. It was a well-known sight to the men and was permanently at his Eighth Army tactical headquarters. Photo courtesy of Wikimedia Commons.

- *If he has friends he will derive strength from their presence and will be anxious not to let them down in battle. He will seek to do his fair share of all tasks which come to his crew or section; he will feel ashamed if he cannot assist his friends in their duties and maintain his place with them in the line.*

- *Morale cannot be good unless men come to have affection for each other; a fellow-feeling must grow up which will result in a spirit of comradeship.*

- *An army is made up of human beings. However much a leader may inspire his men, however perfect the discipline, the morale will be hard and unsympathetic if the warmth of comradeship is not added to it.*

After a fierce battle, members of a British tank crew rename an Italian-German road, "Democracy Lane" in North Africa, 1942. British soldiers kept their spirits up with wry, understated humor. Photo courtesy of the Library of Congress.

- *Comradeship is the spirit of fellow-feeling which grows up between a small group of men who live and work and fight together.*

- *Comradeship is a great antidote to fear because it gives a man friends.*

- *Comradeship is based on affection and trust which, between them, produce an atmosphere of mutual goodwill and feeling of interdependence.*

- *Comradeship is vital to high morale because it surrounds a man with an atmosphere of warmth and strength at the very moment when he is feeling cold and weak.*

- *Comradeship encourages a man's finest instincts and the demands of friendship serve to strengthen him in battle.*

Montgomery (front, left) stands by as King George VI knights Lt.-Gen. J.T. Crocker near the battlefield in October 1944. Montgomery expressed that all men have a "touch of nobility" and needed to maintain high personal standards of behavior in war. Photo courtesy of Wikimedia Commons.

- *Men do not work well together merely because they are disciplined and well-led; they do so because they trust each other and because they are bound together by an affection which is never expressed in words, but shows itself in deeds.*

- *All men have within them a streak of generosity and unselfishness—a touch of nobility—and these qualities will be brought out in their attitude to their friends.*

- *Men must not be moved from unit to unit, or even from platoon to platoon, unless there are good reasons for it, and these reasons must be made clear to the men.*

- *Men learn to have faith in each other and to depend on each other according to the abilities of each.*

Montgomery shakes hands with a paratrooper while inspecting the 6th Airborne Division in March 1944. He believed commanders needed to personally acknowledge their men's achievements in order to boost morale. Photo courtesy of Wikimedia Commons.

4. Self-Respect

- *No man can be said to possess high morale if the quality of self-respect is lacking.*
- *Self-respect implies a determination to maintain personal standards of behavior.*
- *A man who respects himself will neither allow himself to become slovenly nor his quarters dirty. Even in action he will take care to see that his personal appearance suffers as little as possible.*

- *Soldiers must be encouraged to respect themselves at all times and under all conditions.*

- *It is the function of the officer to encourage and instill self-respect. The officer must ruthlessly insist on the maintenance of personal standards. At the same time however he must let his men understand that he appreciates and respects them as human beings.*

- *Efficiency is inseparable from self-respect.*

- *The sense of a good job well done, of a hard task successfully accomplished, is indispensable to good morale.*

- *Men must take pride in their ability to carry out all jobs allotted to them. They must feel that they are good soldiers and therefore of value to other people. Men can be persuaded of this fact by being trusted.*

- *A man who feels he is trusted will feel that he is efficient and he will at once begin to respect himself.*

- *Men who are trusted gain self-confidence. It is the job of the officer to convince his men that he trusts them.*

- *It is true to say that without self-respect good morale is impossible. It is equally true to say that if the standards of leadership, discipline and comradeship are high, the quality of self-respect will also be high.*

ADDITIONAL SPIRIT-STRENGTHENING FACTORS:

- *There are certain contributory factors which powerfully assist morale but do not themselves constitute essential conditions for it.*

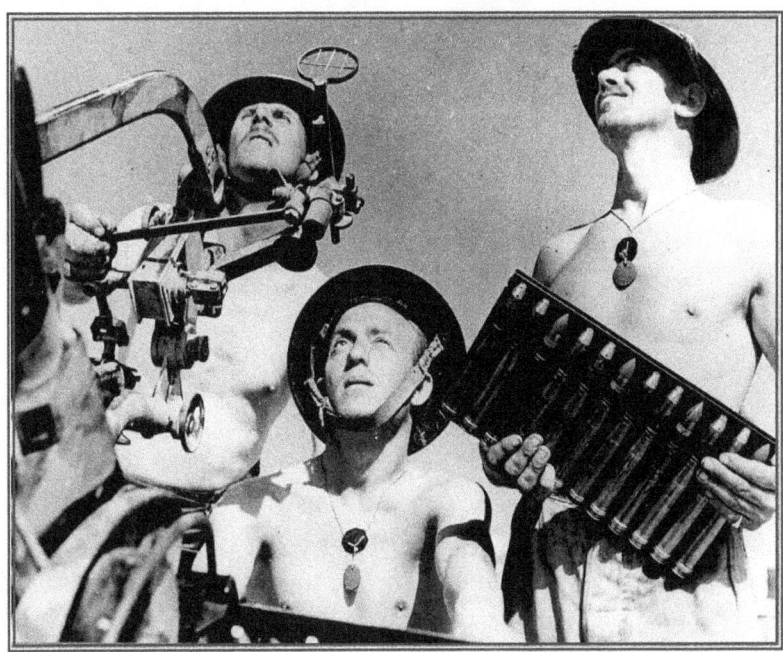

British gunners in June 1941 use a captured Breda anti-aircraft gun to fire at German aircraft within range in Tobruk. "Although besieged by the enemy, Tobruk is by no means inactive. The garrison is busily engaged in preparing of defenses and repelling enemy attacks, and is fairly humming with activity. These give one a peep behind the scenes, and illustrate many of the activities of the Beleaguered Fortress in the desert," noted the British military in the photo caption. Photo courtesy of the Library of Congress.

- *It is possible to have high morale without any of these contributory factors, but it is very difficult. It requires the highest standards of leadership and discipline, and the strongest feelings of comradeship and self-respect.*

- *In the normal case, one or more of these contributory factors must be present. There are many of them, and only a few are considered here.*

1. Devotion to a Cause

- *A democracy cannot oppose the will of the majority of its citizens. The soldier, as a citizen, must therefore be convinced of the rightness of the cause. At least his reaction to the declaration of war must be one of acquiescence, even if this is only passive; he must not be hostile to it.*

- *There are times when a few men will be gripped by a cause and will perform astonishing deeds of heroism to further it. Such men do not require the ordinary bonds of discipline which unite and strengthen others. Their devotion to the cause is in itself all-sufficient. Nor do they require the same kind of leader as has been described. They themselves will all be leaders.*

- *Soldiers are not greatly influenced by cause. There are exceptions, but mainly they fight for reasons which have no connection with freedom or liberty or democracy. Rhetorical statements which assert that the soldier "must know what he fights for and love what he knows" must not be allowed to confuse the issue. The fact is that the soldier, instead of having "a fire in his belly," advances with a cold feeling inside him.*

Montgomery with Canadian Gen. Henry Duncan Graham (H.D.G.) Crerar and Gen. Sir Miles Dempsey stand May 10, 1945 near a Union Jack outside their army headquarters. During his career as a soldier, Montgomery fought for diverse reasons: out of sheer duty in World War I and for a righteous cause during World War II. In World War I, he was not compelled by a cause, but fought out of duty and discipline. During World War II, he had a zealous desire to defend Britain and vanquish the cruel Nazi regime. Throughout his life, Montgomery had deep patriotism. He could relate to all types of soldiers. Photo courtesy of Wikimedia Commons.

- *This is not so in the case of the leaders. Some fight for the same reason as the men; the more intelligent leaders fight because they at heart believe in what they fight for. Such leaders are usually the best in an army and wield the greatest influence.*

- *Numerically, cause is of little importance; but it is a powerful factor because the leaders are greatly influenced by it.*

- *Nothing stated here must be interpreted as minimizing the influence of cause on officers and men who are moved by it. For these few, cause will be a sustaining and strengthening factor and may be of great importance to them. All these remarks apply to soldiers fighting in a war such as the last one [World War II].*

2. Success

- *High morale is possible in defeat but not during a long period of defeat. On such occasions confidence in the leaders will inevitably wane and the first basis will be undermined.*

- *Success will aid good morale by creating confidence in the leader and in the command. This factor requires no enlargement.*

3. Military Tradition

- *The regimental spirit can be a powerful factor in making for good morale.*

- *The more a soldier feels himself to be identified with his regiment, the higher will be his morale if the four essential conditions have been fulfilled.*

Pipers of the British Army entertain Italians in a public square in Sicily after the September 1943 Allied invasion of Italy. Montgomery's British 8th Army participated in the invasion. Photo courtesy of the Library of Congress.

- *It must be realized, however, not only that there can be good morale without strong regimental feelings, but that regiments with a great tradition do not necessarily always produce good battalions.*

- *Regimental spirit is the soldier's pride in the traditions of his regiment and his determination to be worthy of them himself.*

Montgomery inspects the Gordon Highlanders in early 1944. Montgomery held his Scottish troops in high esteem. Photo courtesy of Wikimedia Commons.

• *Nothing but good can result from this spirit which should be constantly encouraged; it is not, however, a basic factor of morale because in the crisis of battle, the majority of the men will not derive encouragement from the glories of the past but will seek aid from the leaders and comrades of the present.*

• *In other words, most men do not fight well because their ancestors fought well at the Battle of Minden two centuries ago, but because their particular platoon or battalion has good leaders, is well-disciplined, and has developed the feelings of comradeship and self-respect among all ranks and on all levels.*

A British soldier rests in a bed inside the remnants of a dwelling during World War I. Photo courtesy of Wikimedia Commons.

- *It is not devotion to some ancient regimental story which steels men in crisis; it is devotion to the comrades who are with them and the leaders who are in front of them.*

- *Regimental tradition assists morale and should be used as a means of developing morale whenever time and circumstances permit. But it is not and cannot be a substitute for morale.*

4. Personal Happiness

- *A man should be happy in the sense that his personal life should be in order.*

A British soldier grins as he receives a lapel flower in 1943 from a woman in Tunisia. Montgomery realized soldiers could not be prevented from fraternizing with civilians in wartime. Instead of trying to stop them, he instead attempted to keep them in top shape for fighting. He occasionally relaxed discipline among his men and ordered that condoms be issued to troops to prevent the spread of venereal disease—a wise and practical decision, although it proved very controversial at the time. Despite his realistic attitude towards soldiers' off-duty behavior, Montgomery would not allow his men to develop a "soft" mindset during war. After conquering Tripoli, Montgomery refused to stay in luxury quarters and instead encamped in the desert outside the city. Alarmed by his men's interest in frolicking in the city, Montgomery quickly ordered his army on the march again to prevent his troops from becoming too comfortable and losing their fighting edge. Photo courtesy of the Library of Congress.

- *Nothing weakens a man more than trouble at home. It encourages him to think of home, and all that it implies, when he should be occupied with the enemy. It turns his mind to peace and his desire to live at the moment when it is necessary for him to steel himself to face the possibility of death.*

- *He must never be allowed to forget that his job is to fight. His function is to kill the enemy and, in so doing, he must expose himself to danger.*

- *A soldier is only a family man after he is a soldier. He must look forward at the enemy and not back towards home.*

- *Anything which weakens his will to fight and expose himself must be considered to lower his morale.*

- *All soldiers do not have their morale affected by home troubles.*

- *Some men thrive on unhappiness and fight all the more fiercely because they hold a secret bitterness within them. Such men are the minority, but they are a large minority. They are a reminder that happiness cannot by itself produce good morale.*

A British paratrooper in Tunisia cleans mud off his boots during World War II. Photo courtesy of the Library of Congress.

5. *Personal Comfort*

- *A soldier's morale will be much improved if the administrative arrangements are good and if he is assured of proper conditions with a reasonable amount of leisure and comfort when he leaves the front.*

- *Hard conditions to teach discipline do not rule out the desirability of good living quarters.*

British soldiers eat lunch in a demolished village during World War I. Forced to fight in gruesome and dirty conditions while deprived of food and rest during World War I, Montgomery attempted to spare his men from having similar experiences. He realized that constant denial of human comforts undermines soldiers' fighting spirit and therefore tried to alleviate his men's sufferings whenever necessary and possible. Photo courtesy of the Library of Congress.

- *The thousand matters embraced in the term "welfare"—from good food to good films—are important because they help maintain a soldier's self-respect and strengthen his confidence in command.*

- *Morale suffers as a result of boredom. Boredom is usually caused by a lack of variety, not by lack of anything to do. Baths, clean clothing, entertainment, newspapers and many little things, which good administration will provide, all kill boredom—but they must be properly organized. Entertainment—if a man misses his meal to see it—is of little value.*

- *The soldier who is well-provided for, who is not disturbed by petty and unnecessary inconveniences, who knows everything possible is being done for him, who is well-clothed and well-fed is a contented soldier.*

- *Letters from home play a great part in morale. There must be no unnecessary delay either in the soldier receiving his letters, or in the letters he writes getting home.*

- *A good delivery system, prompt censoring, and an accurate record for purposes of redirection when a man leaves his unit are all necessary.*

- *Welfare by itself will not produce good morale because it is essentially soft, and morale cannot be good unless it contains a quality of hardness.*

- *Men will endure great hardships if they know why and are convinced of the necessity.*

CHAPTER 4:
APPROACHES TO BATTLE

Montgomery (left) reviews battle plans in Sicily, 1943. Photo courtesy of the Museum of Danish Resistance Photo Archives.

TACTICAL ESSENTIALS:

- *The whole essence of modern tactical methods in battle lies in the following factors:*
    - *— Surprise,*
    - *— Concentration of effort,*
    - *— Cooperation of all arms,*
    - *— Control,*
    - *— Simplicity, and*
    - *— Speed of action.*

The Basic Points of Any Operation:

A) Personal command, control, and good intercommunication.

B) Keep your firepower concentrated under centralized control whenever possible.

C) In all offensive operations, endeavor to hit hard on a narrow front and keep on hitting, penetrate deeply, and then turn outwards, i.e., [following the military principles known as] the Schwerpunkt and the Aufrollen. The momentum of the attack must be kept up at all costs.

D) Fight your brigades as brigades, with definite tasks and clear-cut objectives.

E) Roads and centers of communication are vital. Open them up for yourself. Deny them to the enemy.

F) The tactical importance of high ground is usually very great.

G) Good medical arrangements.

H) Traffic control and careful organization of echelons.

- *All commanders must have complete confidence in the plan.*

- *Throughout the battle area, the whole force must be so well-balanced and poised that there will never be any need to react to enemy thrusts. These can then be disregarded, and the battle forced relentlessly to swing your way.*

Montgomery (center, right) with his team of corps commanders in North Africa (circa 1942). Photo courtesy of Wikimedia Commons.

- *The first requirement of a simple plan is that each component part of the force should have its own task to carry out, and its operations should not be dependent on the success of other formations or units. Once complications creep in, then troubles arise.*

- *The difference between the immediate and deliberate counterattack is entirely one of the time taken to mount the operation—and not of the size of the force involved.*

- *It is necessary to remember that all divisions are different. Some are good at one type of battle, others are good at another type of battle. The art lies in knowing what each division is best at and having the right divisions in the right place at the right time.*

Montgomery (center) rides in a jeep during Army exercises in World War II. Photo courtesy of Wikimedia Commons.

- *It will be exceptional to win a battle without taking certain risks. It requires a nice judgment to decide what risks are legitimate and justifiable, and what risks are definitely not so.*

- *A commander who is not prepared to take a chance and who tries to play for safety on all occasions will never reap the full fruits of victory.*

- *If every unit commander in the army knows what is wanted, then all will fight intelligently and with cohesion.*

- *Before the battle begins, an army commander should assemble all commanders down to the lieutenant-colonel level and explain to them the problem, his intention, his plan and generally how he proposes to fight the battle and how he is going to make it go the way he wants.*

- *The essentials of the battle plan must be known right down through the chain of command and finally down to the rank and file.*

- *Every single soldier must know—before he goes into battle—how the little battle he is fighting fits into the larger picture and how the success of his fighting will influence the battle as a whole. A careful system is necessary to ensure that secrecy is not compromised; commanders in their several grades, and finally the rank and file, must be brought into the picture at the right moments and not so late that they cannot do their jobs properly.*

- *Furthermore, unit commanders will pass on the relevant information to the regimental officers and men, and the whole army goes into battle knowing what is wanted and how it is to be achieved. The resulting effect will be terrific, and nothing will be able to stand against it.*

- *When the troops see that the battle has gone exactly as they were told it would go, the increase in morale and the confidence in the higher command will be immense—and this is a pearl of very great price.*

Montgomery provides instruction to British Army and Royal Navy officers in Port-en-Bessin, France on June 10, 1944 after Operation Aubrey took place a few days earlier during the Normandy Campaign. Photo courtesy of Wikimedia Commons.

Keeping the Initiative:

• *A commander must understand very clearly that without the initiative he cannot win.*

• *When making a plan, it should be remembered that most opponents are at their best if they are allowed to dictate the battle; they are not so good if they are forced to react to your movements and thrusts.*

- *It is necessary to gain quickly and to keep the initiative. Only in this way will the enemy be made to dance to your tune and to react to your thrusts.*

- *Surprise is essential. Strategical surprise may often be difficult, if not impossible, to obtain; but tactical surprise is always possible and must always be given a foremost place in the planning.*

- *The enemy must be forced to dance to your tune all the time.*

- *As the battle develops, the enemy will try to throw you off your balance by counter-thrusts; this must never be allowed.*

- *Throughout the battle area, the whole force must be so well-balanced and poised—and the general layout of the dispositions must be so good—that there will never be any need to have to react to enemy thrusts.*

- *The initiative, once gained, must never be lost; only in this way will the enemy be made to dance to your tune and to react to your thrusts.*

- *If you lose the initiative against a good enemy, you will very soon be made to react to his thrusts. Once this happens, you may well lose the battle.*

- *It is very easy in large-scale operations to lose the initiative, and great energy and drive are required to prevent this from happening.*

CHAPTER 5:
ESSENTIALS OF THE FIGHTING MACHINE

Montgomery at the port of Benghazi during his fight against Rommel in North Africa. Photo courtesy of the Museum of Danish Resistance Photo Archives.

THE MOST VERSATILE FORCE: THE INFANTRY

- *In my mind in modern war it is the infantry soldier who in the end plays the decisive part in the land battle despite of the magnificent part played in battle by aircraft, artillery, and tanks. I do not say this just because I happen to be an infantry soldier myself; I believe it to be true.*

- *The infantry is the most versatile of all the arms; it can operate in any weather, in any type of ground (mountains, forests, jungle, swamps, and desert).*

- *The infantry soldier remains in battle day and night—with little rest and without adequate sleep. He can use very expressive language about the way he has to bear the main burden in battle, but he does it! I salute him.*

- *You cannot have a good army without good infantry.*

- *The Infantry Division has many roles in battle, but by far the most important is its attack upon, and break into, a main enemy-defensive area.*

- *This will entail a hard slogging match with great calls for stamina, fortitude and endurance on the part of all ranks of all arms in the division.*

- *Active and aggressive infantry patrolling and sniping will ensure that he retains the initiative. This will also provide important sources of information about enemy strength, dispositions, defenses, and morale.*

Flexible Firepower: Tanks

- *A tank is an armored vehicle designed to carry about firepower. This definition, once understood, simplifies the problem of the use of armor on the battlefield.*

- *The weight of any tank should not exceed about 45 tons.*

Tanks from the British 8th Army in Tripoli travel in rough terrain during World War II. Photo courtesy of the Library of Congress.

- *The positioning of the armor on a tank is important. Experience in the field has proved that the percentage of hits is higher on the sides than on the front of a tank. Also there are, on the average, more hits on the hull than on the turret. The reason is because usually enemy guns are lying in wait and do the most damage.*

- *As a result of practical experience in battle, it is important to study the slope and thickness of the armor to obtain the maximum basic protection over all the vulnerable places that have become disclosed.*

- *No plan for using an armored division will be sound if it does not fully exploit these main characteristics of an armored division below:*

    A) Its armor,

    B) Its firepower, and

    C) Its mobility.

- *The armor is most effective when employed concentrated. A mass of armor, particularly in the enemy's rear, has a great moral effect.*

- *The armored brigade possesses tremendous firepower. It is vital that this should not develop into fire without movement. Tanks must not be used as artillery. Once this practice is allowed to develop, the offensive spirit and will to get forward—which are so vital for success—will be lost.*

- *Although possessing great mobility, tanks are generally sensitive to ground. The terrain must be carefully studied to ensure that the armored division is not unexpectedly prevented from exploiting its main characteristics.*

British troops on a General Grant tank plough through a wet and muddy road during the Allied advance in Libya during World War II. At the height of the Allied pursuit of the Axis forces, heavy rain fell in the Western Desert and turned the sand and dust into mud and flooded low-lying parts. Photo courtesy of the Library of Congress.

- *The commander of an armored division must be prepared to regroup constantly to meet changing circumstances and varying types of terrain. He must therefore look well ahead at all times.*

- *The division must be so flexible as to permit constant regrouping. The machinery for carrying this out quickly must exist.*

- *Intimate air support is of great importance in offensive operations by an armored vision.*

- *Night advances by an armored division require previous training and practice.*

- *Tanks alone are never the answer to any problem. Success will be obtained only by the most intimate cooperation of all arms in a division.*

AIR POWER:

• *Winning the air battle is a prerequisite to military success.*

• *Any officer who aspires to hold high command in war must understand clearly certain principles regarding the use of air power.*

• *All modern military operations are in fact combined Army/Air operations. It is essential that the staffs of both services work together from the outset on a joint basis in planning operations—with complete mutual understanding and confidence.*

U.S. Air Force General Curtis LeMay, SAC Commander, (left) and Field Marshal Viscount Montgomery, Deputy Supreme Allied Commander in Europe, discuss a model of the Strategic Air Command's latest heavy bomber—the B-52—during Montgomery's visit to SAC headquarters in November 1954. Montgomery was briefed on SAC's worldwide operations and capabilities. Photo courtesy of the Library of Congress.

• *The greatest asset of air power is its flexibility. The concentrated use of the air striking force is a battle-winning factor of the first importance.*

Air-landing troops board a Hotspur glider during a training exercise in Britain, November 1942. Montgomery was appointed Colonel Commandant of the Parachute Regiment in 1944. Photo courtesy of Wikimedia Commons.

- *Military commanders must have a working knowledge of the capacity and limitations of air weapons.*

- *The military commander cannot make his plans or frame reasonable requests without some knowledge of his subject—just as he must know enough of his artillery weapons and other supporting arms.*

- *No commander can form an adequate appreciation of the potentialities of supporting air action without some knowledge of the characteristics of the different types of aircraft.*

- *The commander of an army in the field must deal in planning air operations with one Air Force commander and one only.*

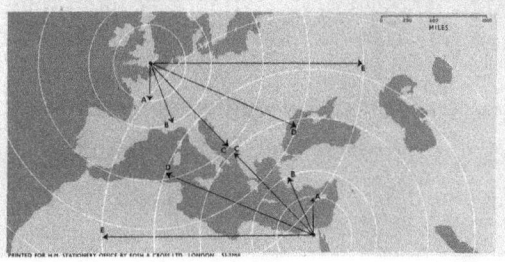

British poster from World War II about aircraft in the Royal Air Force. Public domain image.

- *The choice of weapons to be used to engage a certain type of target must always remain an Air Force decision.*

- *Control of the available air power must be centralized and command must be exercised through Air Force channels. The soldier must not expect or wish to exercise direct command over air-striking forces.*

- *Military commanders and staffs should take every opportunity for gaining knowledge of the chief characteristics of the various aircraft which may operate on their behalf and as a result of their request.*

Fundamentals of Air Power:

- *It is necessary to win the air battle before embarking on the land battle. If this is not done, then operations on land will be conducted at a great disadvantage.*

- *It is impossible to conduct successful offensive operations on land against an enemy with a superior Air Force, other things being equal. The enemy's Air Force must be subdued before the land offensive is launched.*

- *Fire support from the air is dependent on the weather. The overall plan must aim at winning the battle with the fire support available from the ground—especially when time is important and you cannot wait. If the air support becomes available, it is a good bonus and enables you to win more quickly and with fewer casualties.*

- *There will be occasions when fire support from the air is essential for success. You then require good weather and must wait for it.*

- *A retreating enemy offers the most favorable targets to air attack. The greater the pressure applied on the ground, the more disorganized the retreat, and the greater the opportunities for the Air Forces to inflict punishment.*

- *The moral effect of air action is very great and is out of proportion to the material damage inflicted. In the reverse direction, the sight and sound of our own Air Forces operating against the enemy have an equally satisfactory effect on our own troops. A combination of the two has a profound influence on the most important single factor in war—morale.*

- *During pauses while the land battle is being built up, the Air Forces must be very active. They should interfere with enemy movement, destroy communications, disrupt his supply organization, and generally carry on the battle while the Army is preparing to deliver its main blow.*

The Douglas A-20 (Havoc) light bomber, called the Boston by the British, was used by both American and British air forces during World War II. Photo courtesy of the Library of Congress.

Information from the Air:

- *Every aircraft which flies over the battle area is a potential source of information.*

- *Reliable and timely information of enemy dispositions and movements is vital to the combined plan and conduct of operations.*

- *The air is an important source, but only one of many sources from which the general enemy situation is built up.*

- *Information from the air is of primary interest to intelligence, and the closest links must exist between air reconnaissance and military Intelligence sections.*

Targets for Air Attack:

- *It is quite impossible to provide rigid rules by which targets can be defined precisely as suitable or unsuitable.*

- *Ideally, a target offered for air attack should be:*
    - *—Readily recognizable from the air,*
    - *—Within the accuracy limits of the air weapon,*
    - *—Vulnerable if hit, and*
    - *—Beyond the capacity of ground weapons.*

- *The importance of the objective in relation to the operational plan and suitability of the target will frequently conflict.*

Bombs lie on an Allied airfield ready to be loaded into Royal Air Force planes during World War II. The British bombers of the Mediterranean Allied Air Forces worked with Americans to strangle German supply lines feeding Nazi troops on the Anzio, Cassino, and Eighth Army fronts. French airmen also participated in the operation. Photo courtesy of the Library of Congress.

- *It must be understood by all concerned that:*

    —Operational priorities may overrule the unsuitability of the target as an air target.

    —In battle, it is seldom possible to reach the specification agreed for the ideal air target.

    —In many instances, intelligence information inevitably is limited to deduction and cannot be confirmed.

- *Target unsuitability can sometimes be overcome by increasing the effort. Superficially this may appear uneconomical but, in fact, circumstances may make it the exact reverse.*

- *Suitability of the target may be judged not by the range of ground weapons, but by their capacity to reduce it.*

- *New weapons and new tactics will continue to increase the scope of air attack.*

- *The timely use of airborne forces may often play a decisive part in land operations.*

- *If the weather conditions are uncertain, the commander must decide how long he is prepared to wait for suitable weather if conditions on "D-day" are unfavorable.*

- *To ensure efficiency in planning air-support operations, it is essential for Army and Air Force commanders and their staffs to understand the requirements, capacity, and limitations of each other's service.*

- *To reach proper standards in these things, it is necessary for both Army and Air Force services to recognize their responsibilities in a practical manner and to introduce appropriate instruction at all levels and at all stages in training.*

- *Technical developments in the air weapon continue swiftly and their possibilities are bounded only by the imagination.*

Montgomery (left) confers with Air Marshal Sir Arthur Coningham (center) and Gen. Sir Miles Dempsey in Germany in 1945 shortly before Montgomery gave the historic order to begin the direct assault crossing of the Rhine. Montgomery's breakthrough crossing of the famous Rhine was a huge blow to German morale. Photo courtesy of Wikimedia Commons.

COOPERATING WITH OTHER MILITARY BRANCHES:

- *Successful battle operations depend on the intimate cooperation of all arms; no one arm—alone and unaided—can do any good in battle.*
- *Success is bound to be conditioned by many factors, of which I consider the following the most important:*
   - *—The degree of knowledge possessed by each service of the other's task, their capacity, and their limitations.*
   - *—The degree of mutual trust and honesty of motive reached between the services.*
- *All modern military operations are combined service operations.*

- *Independent services must operate smoothly and efficiently in what is fundamentally a common task. This automatically implies a process of negotiation rather than authority—and a satisfactory solution is no easy matter.*

- *Satisfactory relationships are not brought about by subscribing them in theory or by giving formal approval to well-meaning principles. They call for hard work and genuine effort on both sides and a constant exchange of views.*

- *All services have something to contribute to joint problems; and in all our dealings with one another, we must be honest and say what we think without being hampered by too narrow a division of responsibilities or by out-of-date formulas.*

- *This sort of approach to common problems is essential to good relations, and genuine progress and improvement.*

Montgomery meets with military officials in Denmark, 1945. Photo courtesy of the Museum of Danish Resistance Photo Archives.

Visits:

• One of the best methods by which services can get to know one another better is by an interchange of visits, and whenever practicable every effort should be made to make arrangements of this sort.

• Visits are of value only when they are properly organized and conducted. Without this, they do more harm than good and merely waste time.

• A properly organized series of visits over a period under active-service conditions helps considerable towards a mutual understanding of one another's difficulties and the conditions under which we each live and fight.

CHAPTER 6:
BATTLE MANAGEMENT

Montgomery oversees battle near Tripoli in January 1943. Photo courtesy of Wikimedia Commons.

THE STAGE MANAGEMENT OF BATTLE:

- *To be successful in battle, the fighting machine must be so set in motion that it can develop its maximum power rapidly, and troops must then be launched into battle properly.*

- *It follows that what may be called "the stage management of the battle" must be first-class.*

- *The plan of the battle must be made by the commander and not by his staff.*

- *In his plan of battle, the commander must give careful thought to the correct grouping of his divisions, his armor, his artillery, and other resources.*

- *He cannot decide on this grouping until the problem has emerged, and he has decided how he will solve it. Then, he must group his divisions, his armor, and his artillery corps—who have to fight the tactical battle—so they are suitably composed for their respective tasks. As the battle proceeds, he may frequently regroup.*

- *Skill in grouping—and in quick regrouping to meet the changing tactical situation—plays a large part in successful battle operations. It is a great art and requires much study before proficiency is attained.*

- *No good results will be obtained by splitting up divisions. Such action affects morale adversely. Nor can a division conduct effective offensive operations against even moderate opposition in good delaying country if strung out on a wide front since it cannot then develop its full fighting power.*

- *Having made his plans, the commander will have much detailed work to do before the operation is launched. This detailed work is the province of the staff.*

- *Once the battle has started, everything that passes between the higher commander and his subordinate generals should be verbal. If this is not always possible because there is no telephone or distances are too great for a personal visit, then written messages may be necessary. As the commander has trained his subordinates to work on his verbal orders, and mutual confidence in dealing in this way has been established, all such messages should be drafted by the commander himself to ensure they will convey to the recipient exactly what the commander wishes.*

- *In mobile operations when it is necessary to strike hard, deep and to penetrate quickly into the enemy country, divisions should operate on narrow fronts on main axes of advance. If the enemy is widely dispersed in his endeavors to stem the advance, he will not be able to hold these "divisional thrusts."*

- *The time may come when the enemy will recover his balance and stabilize the battle on some rear position. When this happens, two or more thrusts should be inclined towards each other to converge on the vital or key locality in the enemy position.*

An April 12, 1945 photo shows Montgomery (center) with his team of Liaison Officers, who were tasked with stealthily gathering information about their own troops in the field and reporting back to him. Montgomery pioneered this system in World War I. Like many commanders, he tended to promote people he trusted; thus, many of these officers were former aides-de-camp. He chose liaison officers with outgoing and friendly personalities to avoid suspicion when gathering information. Left to right: (back row) Maj. Sharp, Maj. Brisk, Maj. J. Poston, Maj. Frary; (front row) Maj. C. Sweeny, Maj. Harden, Montgomery, Maj. Earle and Maj. Howard. Public domain photo.

GATHERING INFORMATION ON ONE'S OWN TROOPS:

• *An essential feature in direct and personal command is the system of liaison officers.*

• *A commander in the field requires a team of highly trained liaison officers to tour the battle area and visit subordinates to bring back each night an accurate and vivid picture of what is going on.*

• *This skilled work is of great importance and first-class officers are required for it. They must be young, active, fearless, very tough and hard, and mentally alert; they must have an attractive personality which will make them welcomed by subordinates of all grades. They must have that character which will enable them to obtain their information without creating suspicion at any level.*

- *Obviously, they must have a good military knowledge. They must also have a high sense of duty to put the efficient performance of the task before everything else.*

- *The final essential is that they must be a team together with the commander who employs them. They are the eyes and ears of their commander—going everywhere and seeing everything. They must know exactly what their commander wants to know and what he does NOT want to know, and be able to give him that information clearly and quickly, omitting all irrelevant detail.*

- *Only the very best type of young officer is suitable for this work and the team, once formed, must be changed as seldom as possible.*

- *This system enables the commander to be in the closest touch with the operational and battle situation. Therefore, he can adjust his dispositions to the battle as it develops and take quick advantage of some favorable situation, or can take steps to prevent enemy interference with his own plans should any such interference seem likely to mature.*

- *Liaison officers take their orders personally from the commander and give their reports verbally direct to him. This "hot information" should be given without delay, and the commander must see the team every night at HQ when they get back from their missions.*

A British soldier helps a comrade take a bath next to a heavy gun in World War II. Photo courtesy of the Library of Congress.

Increasing Morale in Newcomers during Battle:

- *Men will often arrive after a long and unpleasant journey to join a battalion of which they have never heard—in other words, "browned off." They must not go into battle till this has been put right.*

- *The "stranger" must become a "member of the team." He will be taught among other things:*

    — *Who his new unit is,*

    — *What it has done,*

    — *What his unit and divisional signs are, and*

    — *Who are the personalities of this unit, his brigade*

and his division, and be made to feel that they are his.

• *His equipment must be checked over and made up if necessary, and he must be medically examined.*

• *When the time is opportune, he must be delivered to his new unit. On no account should he join at a time when the stress of battle makes his unit too busy to "sort him out" and so parks him in "B" echelon.*

• *To obtain the best value from a reinforcement, he must be first welcomed into his new team, then immediately absorbed into it.*

• *New and untried troops must be introduced to battle carefully and gradually, with no failures in the initial ventures. A start should be made will small raids, then big-scale raids, leading up gradually to unit and brigade operations.*

• *Great and lasting harm can be done to morale by launching new units into operations for which they are not ready or trained, and which are therefore likely to end in failure.*

With the 8th Army in Tripoli, the Royal Air Force Rescue launched a speedboat to action from a desert port on the Mediterranean during World War II. Photo courtesy of the Library of Congress.

• *When new units and formations are introduced to battle there must be no failures.*

Overseas Campaigns:

• *The early appointment of the commanders who will carry out the operation is essential. They should be appointed before the planning begins.*

• *Any overseas campaign will involve the closest cooperation between the Navy, the Army and the Air Forces.*

• *The Navy has got to take the Army across the seas, and this requires good beaches for landing. The Army when on shore cannot be maintained indefinitely over open beaches, but requires a good port very early. The Air Forces require good airfields.*

The P-40 single engine fighter plane was used by the British in various models as the Kittyhawk, Tomahawk and Warhawk during World War II. Photo courtesy of the Library of Congress.

- *But the overall plan of the battle must not be built up solely on the need to acquire quickly good beaches, good ports, and good airfields.*

- *The whole question is the conduct of offensive operations in an enemy country with the object of destroying the enemy's armed forces and occupying his territory. The Army has got to carry out this task, and no other service can do it.*

- *Therefore, the first need is to decide how you want the operations on land to be developed so that the object can be successfully attained in the simplest and quickest way.*

- *It is then for the Navy to say whether the Army can be put on shore in such a way that the land battle can be developed in the required manner. It is for the Air Forces to say whether this will suit the air plan.*

Montgomery (second from left) reviews a plan with senior officers in 1941. Photo courtesy of Wikimedia Commons.

• *And so the plan is built up. Some compromise may be necessary, but eventually an agreed plan will emerge. The beaches, ports, and airfields then become objectives in the general plan of battle.*

ORGANIZATION:

• *The wise commander will keep a secure hold over the basic operational aspect of the battle and will not let it be taken away from him by his staff.*

• *The commander has the responsibility that the organization, training, and handling of his administrative staff and units are as sound as that of his fighting units; and his administrative arrangements are always in line with his tactical plan. This responsibility cannot be delegated.*

Tanks of Montgomery's Eighth Army line up in Tripoli in 1943. Army administration involves the supply, transport, and distribution of essential items including food, equipment, weapons, ammunition and vehicles to fighting troops. Although many commanders in history struggled to manage their administration while commanding in the field, Montgomery proved adept at it. Photo courtesy of Wikimedia Commons.

- *Constant changes cause confusion and loss of efficiency—both tactically and administratively.*

- *Good administration is economical.*

- *Economy does not mean an all-round cut of 10%. It means the avoidance of waste.*

- *Resources should be provided on an "as required for the*

A British Sherman tank crew handles ammunition in Germany in March 1945. Photo courtesy of Wikimedia Commons.

operations" basis. The ruling factor must be, "Is this item needed?" NOT "Am I entitled to this item?"

• *Administration is a mass of detail. The commander must not get immersed in this. It is the responsibility of his staff.*

• *Administrative problems will be many and varied. A too-rigid organization will never solve varied problems. Elasticity is required and must be developed. This will be achieved only if administrative communications are adequate.*

- *Administrative plans are always subject to limitations—either of transport, supplies, time, or distance. Therefore, the best possible use must be made of everything, and conflicting requirements carefully balanced against each other.*

- *All demands must be accurate, timely, and the result of careful calculations. To over-demand means either that you go short of something else, or somebody goes short of what you have over-demanded—or often both.*

- *The capture of stores by the enemy may present him with the only means by which he can make an effective counterstroke. As a general guide, it may be stated that forward reserves should be held to a minimum.*

- *The results of under-insurance inevitably become apparent to all, whereas the crime of over-insurance does not become apparent, and may indeed lead to a feeling of satisfaction that supplies of all sorts are plentiful and the administrative arrangements are excellent. Of course, the reverse is the truth—as any form of over-insurance must inevitably cramp the commander in his operations and prevent him from making full use of his opportunities.*

- *A good staff officer must—in addition to the qualities of foresight, accuracy and attention to detail—possess the human touch. He must have a thorough knowledge of the Army and be able to realize the effect of his orders on the eventual recipient. Bad staff work produces dissatisfaction among the men, and dissatisfaction leads to loss of morale.*

A truckload of British 8th Army soldiers wait as their vehicle refuels at a depot. Gasoline canisters are spaced evenly across the area as troops prepare for the Tripoli campaign in North Africa during World War II. Photo courtesy of the Library of Congress.

- *It is of prime importance that a commander choose a chief administrative officer who can nicely calculate his risks in the light of the probable course of events, and who has sufficient detailed experience to know and check potential sources of over-insurance on the part of the staff.*

- *Early consideration of all possibilities by heads of branches and services is essential as administrative plans have to be laid a long time in advance.*

- *No situation should be allowed to arise to which some thought has not already been given.*

- *Shipping in particular—and provision of adequate reinforcements—are matters that take time to arrange and cannot be put into operation at the last hour.*

Administrative Demands in Battle:

- *The commander must place full confidence in his chief administrative officer's advice and, once having agreed the plan, must keep within the limits laid down.*
- *Occasions may arise when additional risks should be deliberately taken. On such occasions it will almost invariably be found that for a short period an extra effort can be made and the scope of the original administrative plan exceeded, but this will inevitably lead to some temporary administrative dislocation and require a pause for reorganization.*
- *The commander must not make the successful outcome of a special call the basis for overstraining continually the administrative machine. If this is done, serious dislocation will ensue and a tendency to over-insurance and lack of confidence will be encouraged.*
- *The administrative system must be flexible, and the administrative staff must be prepared to improvise.*
- *It is only by means of close cooperation that forward movement and protection of the large and vulnerable administrative tail of the division can be ensured.*
- *Complete confidence between the administrative staffs of higher and lower formations is essential.*

CHAPTER 7:
ATTACKING & DEFENDING

Montgomery reviews troops. Photo courtesy of the Museum of Danish Resistance Photo Archives.

THE SET-PIECE ATTACK:

- *The first aim of the attack is to capture ground, the second to hold it.*

- *The basic points of the Set-Piece Attack:*
    1. *The attack must be organized in depth,*
    2. *The start line must be secure,*
    3. *The attack must be "seen in" by fire,*
    4. *Assaulting infantry and tanks must keep close up to the fire,*

> 5. Supporting weapons must get forward quickly, and
>
> 6. The impetus of the advance must not be allowed to die down.

- *Depth in attack is necessary to two main reasons:*

  > 1. To maintain the momentum of the attack. (Fresh troops should be ready to go though even if the first attacking troops have not got all their objectives. An attack should always aim at deep penetration to overrun enemy mortar and gun positions.)
  >
  > 2. To mop up in the wake of the assaulting infantry and tanks. (This mopping up must follow very quickly and be very thorough.)

- *Infantry and tanks must get right up to their supporting fire and ready to go in immediately as it lifts, and so overrun the defense before the latter can get its "second wind."*

- *If the start line is not secure, the whole plan of attack will be in danger of complete failure. A "dog fight" may develop for the start line, and the fire plan will not be designed to cope with this. The deployment of the assaulting troops will probably be seriously delayed and interfered with.*

- *The assaulting troops must be assisted forward by all available support from artillery, mortars, machine guns, and air. This fire will aim at covering all known or likely enemy localities, including those which can support the defender by fire from flank and rear.*

- *The defense must be shaken and stunned by fire or bombing. The battlefield must be isolated. The importance of counter battery fire must never be overlooked.*

British tanks under Montgomery's command roar through dust clouds on the offensive during the Battle of El Alamein, 1942. Photo courtesy of Wikimedia Commons.

- *Great firepower is useless unless the assaulting troops are able to take quick advantage of it.*

- *It is only by continuous pressure that a breakthrough can be achieved. Once the attack has started, the enemy must be given no respite in which to organize or collect reserves.*

- *Fresh formations should be constantly moving up ready to move through the leading troops so the momentum of the attack may be maintained relentlessly by day and night.*

- *In many attacks, it will be necessary to breach minefields, antitank ditches, and concrete defenses... Previous practice and training is required, particularly from the point-of-view of command and control.*

- *When attacking through built-up areas or when concrete pillboxes exist, flame-throwing tanks are of great value.*

Maintaining the Momentum of an Armored Attack:

Closeup of a British 8th Army tank used in the Tripoli campaign. Photo courtesy of the Library of Congress.

- *The first objective of the armored division must be the capture of some place of high ground, the loss of which will make further resistance by the enemy impossible in his present position.*

- *No matter what success the assaulting formation may achieve, the armored division must follow closely on its heels ready to exploit any opening which is offered.*

- *To maintain the momentum of the attack, it is essential that the armor, infantry, and guns move forward without any delay. Immediately as the armored division is ordered to advance, the assaulting formations must clear all routes to allow the armored division an uninterrupted passage.*

British soldiers man a machine gun position near a tank in Italy, 1944. Photo courtesy of Wikimedia Commons.

• *After the enemy has been driven from his defensive positions, the battle is now mobile. In order to conduct a successful withdrawal, the enemy will have to make use of roads and centers of communication, some of which may be defiles. These must be denied to the enemy or captured, thus turning his withdrawal into a rout.*

Principles for Infantry and Tank Cooperation:

• *Infantry and tanks must "marry up" early.*

• *Reconnaissance of ground and obstacles (for tanks) is necessary before every attack.*

—Bernard Montgomery's Art of War—

- *The formation adopted will depend on the ground.*
- *The tank is an offensive assaulting weapon.*
- *Good inter-communication between tanks and infantry is essential.*
- *Tanks must have time to replenish ammunition and fuel, and for maintenance.*

Rules for Assaulting and Exploiting a Water Obstacle:

A Sherman tank disembarks from a landing craft and drives ashore at Anzio during Operation Shingle in January 1944. Photo courtesy of Wikimedia Commons.

- *Detailed reconnaissance of the obstacle is necessary to select infantry crossing places and bridging sites.*

Tanks under Montgomery's command cross the Rhine near Wesel in March 1945. Montgomery successfully broke through German defenses on the formidable Rhine during Operation Plunder using fierce Scots "River Sweeper" soldiers to launch the first wave of the assault. Men of Scotland's famous Black Watch Battalion were the first British soldiers to cross Germany's national river in a historic attack. Photo courtesy of Wikimedia Commons.

- *Rivers and canals form good delaying obstacles, and the enemy will attempt to make full use of them. The first aim of a commander will be to seize a bridge intact, and so press and harry the enemy so that he is unable to make an effective stand.*

- *If this fails, a deliberate attack must be organized to cross the obstacle, form a bridgehead, and build a bridge.*

- *Every effort must be made to capture a bridge intact by a "coup de main."*

- *An armored division cannot undertake a major operation of this type. This is basically an infantry operation. However, from the time that it approaches a water obstacle, it must do everything possible to achieve a crossing.*

  — *The infantry crossing should be on a wide front. Reserves should be centrally placed so they can be switched to the most successful crossing place.*

  — *The infantry bridgehead must be deep enough to cover the bridging sites from aimed small-arms fire.*

  — *Antitank guns must be got across early.*

  — *There must be careful organization, and control of all troops and stores crossing the river.*

  — *The bridgehead must be expanded as soon as possible after the completion of the bridge.*

- *The enemy will counterattack with all his available reserves to drive the infantry back before the bridge can be built. As in every other form of attack, it is of vital importance to get supporting arms—particularly antitank guns—across the obstacle as early as possible.*

- *Bold action on arrival at the water obstacle may well avoid the need for an assault operation. If a deliberate assault is necessary, the armored division must assist by carrying out wide reconnaissance, securing the flanks, and supporting the assault by fire from tanks in hull-down positions on the near bank.*

An American mortar crew bombards the enemy in attempt to establish a Rhine beachhead in March 1945. Photo courtesy of Wikimedia Commons.

- *On arriving at a water obstacle, the division must endeavor to seize and hold any bridges which the enemy has not destroyed. A success at one point must be reinforced immediately, and a strong bridgehead established.*

- *Even if all the bridges are destroyed, every effort must be made to secure a small bridgehead supported initially by fire from the near bank until supporting weapons can be rafted across.*

- *Close defense of a water obstacle along its whole length requires so many troops that it will seldom be possible. Reconnaissance must be carried out to find suitable infantry crossing places that are not closely defended.*

- *A fire plan should be prepared so that as soon as the bridge is completed a really determined outward thrust can be made with armor and infantry.*

- *As in every other form of attack, it is of vital importance to get supporting arms across the obstacle as early as possible.*

- *Very careful control of the routes leading up to the bridge must be arranged.*

American troops land on the beachheads of Normandy during the famed Allied "D-Day" invasion of June 6, 1944. Montgomery commanded invading Allied forces during and after the landings. Photo courtesy of Wikimedia Commons.

A bird's-eye view of Allied troops June 6, 1944 on the beaches of Normandy. Montgomery assisted in successfully planning the invasion. In 1944, Montgomery commanded no less than 2 million men, including American troops. Photo courtesy of the Library of Congress.

ASSAULTING AN ENEMY-HELD COAST:
• *The commander must retain the initiative during the early fluid fighting. The enemy will then be forced to conform, and to use up his reserves piecemeal in stopping gaps, rather than concentrated against the beaches.*

The Basic Points

1) *Speed and order of landing are the first essentials.*

2) *Landing beaches must be freed from observed small-arms fire.*

3) Landing beaches must be joined up.

4) Armored columns must be pushed ahead quickly inland.

5) Keep the initiative.

- *It is most important to get enough infantry and supporting arms quickly onshore to withstand immediate counterattack. This will depend largely upon sound preliminary planning.*

- *There will never be enough space for all you would like. Decide what is essential.*

- *The rapid capture of the first infantry beachhead, and complete mopping up of all the enemy in this area, must aim to free the landing places from small-arms fire. The next step is to deny the enemy observation of them for his mortars and artillery.*

- *The infantry division may land on a wide front at several beaches. Once these are firmly joined up, the enemy's chance of isolating them, and dealing with them in detail, has gone.*

- *Boldness in pushing inland with infantry and tanks to seize important ground will prevent the enemy from developing his attack against the landing places.*

A British Bren gun carrier passes by destroyed enemy material in North Africa (circa 1942). Montgomery's victory at the Battle of El Alamein in 1942 marked the first time that the elusive Field Marshal Rommel experienced a decisive defeat in open combat. Photo courtesy of Wikimedia Commons.

Pursuing a Withdrawing Enemy Force:

The Basic Points

>1) Threaten the enemy flanks and always be prepared to bypass the leading troops.
>
>2) Keep touch.
>
>3) Smash through on a narrow front.
>
>4) All available artillery should be well forward.
>
>5) Drive hard through to seize important communication centers and thus cause the enemy resistance to disintegrate.
>
>6) Keep up the pressure by day and if possible by night.
>
>7) Watch the administrative situation.

British paratroopers advance in March 1945 through Germany in pursuit of enemy forces following airborne landings east of the Rhine. Photo courtesy of Wikimedia Commons.

- *The object of the enemy will be to impose delay—to prevent a breakthrough from disorganizing his withdrawal—and gain time to organize his next main defensive area. He will attempt to do this with mines, demolitions, and fire.*

- *Pressure must be ruthlessly maintained. The commander should state at the beginning of the operation what size of enemy pockets may be bypassed by the leading troops.*

- *A retreating enemy force is always frightened of encirclement and is particularly sensitive to any outflanking movement. To threaten these, it is necessary first to find them and then probe hard on a wide front and discover the weak spots.*

British sappers clear mines from beach at Normandy in July 1944. The Germans had fortified the beaches under the command of Rommel, who developed innovative ways of using mines to delay pursuing forces. Due to his experiences fighting Rommel, Montgomery learned much about overcoming hazards and deceptions created by a fierce enemy in retreat. Photo courtesy of Wikimedia Commons.

- *Though all centers of enemy resistance must eventually be reduced to open the axis, the leading troops should always attempt to bypass these localities and leave the clearing up to be done by subsequent echelons of troops.*

- *It is vital to ensure that the enemy is not allowed to slip away behind cover of small, but determined rear guards. Vigorous patrolling should be coupled if necessary with a series of carefully staged attacks. These will give early information of his intention.*

- *All available firepower should be concentrated to smash though on one axis, and the resources concentrated to open up the other axis.*

British anti-aircraft guns in action near Tobruk during a night raid in North Africa. Photo courtesy of the Library of Congress.

- *A succession of rapid hard-hitting punches will be necessary, first with one battalion, then with another.*

- *The launching of the second attack should not be abandoned just because the first one has not fully succeeded.*

Defense:

- *All defense is temporary and is the prelude to offensive action from the ground defended or from elsewhere.*

A camouflaged British sniper wearing a German camouflage smock demonstrates new techniques in concealment at a sniper school in France, July 1944. During World War II, the Germans developed highly effective camouflage patterns based on natural colors and light patterns. Many of these concealment techniques were initially fielded by sinister Waffen S.S. troops. Photo courtesy of Wikimedia Commons.

The Basic Points

1) *Localities must be concealed. This is the first consideration.*

2) *A screen of forward troops must cover the main and vital centers.*

3) *Control of artillery must be centralized.*

4) *Localities must be organized all-around defense.*

5) *Troops including armor must be available for immediate and deliberate counterattack.*

6) *Obstacles must be covered from enemy reconnaissance.*

Members of an American patrol use white bed sheets for camouflage during the Battle of the Bulge in December 1944. Photo courtesy of Wikimedia Commons.

- *In battle, attack and defense alternate. It is impossible to attack all day and all night—and when resting or regrouping for another phase—a division must quickly assume such defensive measures as are suitable.*

- *All defense must be aggressive and threatening—both to mislead the enemy and retain the high morale of one's own troops.*

- *A defended area is a firm base from which to develop offensive action—by fire—or by movement—or both, as suitable to the conditions of the moment.*

- *Vigorous offensive patrolling and sniping, coupled with the offensive use of supporting artillery, mortars, and machine guns will enable the commander to retain the initiative and keep the enemy at arm's length.*

- *Concealments of all troops in a defensive area requires great care in the selection and preparation of localities. It may entail restrictions of movement by day. This is of first importance since without a knowledge of the defensive layout, any enemy attack is made extremely difficult.*

- *To achieve concealment, infantry localities will often be best sited on reverse slopes.*

- *An outpost screen of automatics and snipers (with artillery) must cover the main defensive area and prevent enemy reconnaissance.*

- *All localities must be organized for all-round defense so that they can hold out even if the enemy has overrun flanking localities and penetrated to the rear.*

CHAPTER 8:
THOUGHTS ON NUCLEAR WARFARE

Montgomery with war correspondents in 1941. Photo courtesy of Wikimedia Commons.

- *War cannot abolish itself. In the simplest terms, nuclear weapons have given mankind a choice between either abolishing war or being abolished by it.*

An atomic bomb explodes in August 1945 over Nagasaki, Japan. Montgomery was thoroughly shaken by the prospects of atomic warfare and mutually assured destruction. He wrote that he believed the U.S. atomic bombings of Japan in World War II were unnecessary and immoral. He had studied the war in the Pacific in detail and concluded that Japan was already on the brink of surrender due to extreme conventional bombing damage and political pressure. Montgomery remained morally opposed to nuclear war for the rest of his life. Photo courtesy of the Library of Congress.

- *Our aim must be to prevent war. The prospect of winning or losing is not a profitable subject. We must find another court of last resort for adjusting political differences.*

Nagasaki, Japan after atomic bombing in 1945. Following a detailed study of nuclear war, Montgomery believed peace was the best answer to global conflict. Although Montgomery, a career soldier, was not a pacifist, he witnessed many atrocities during two world wars and believed further destruction would cause the great sacrifices of soldiers and civilians to have been made in vain. Photo courtesy of the Library of Congress.

- *I have been studying nuclear war for a considerable time, and I have come to the conclusion that man will have it within his power in the future to destroy himself and every living thing in this planet. I do not believe this to be man's destiny. But we must face the facts now, or it will be.*

- *You may say: 'How can we prevent war? Man has been warring since the dawn of history. Why should he change his ways now? Man will, as far as one can foretell, always make war unless there is some powerful deterrent to prevent him.' Here lies the key to our problem.*

U.S. troops watch an atomic bomb test in 1951. Photo courtesy of the Library of Congress.

- *The banning of nuclear weapons will not give us peace. We will get lasting peace only by having the nuclear deterrent, as no nation will risk its own utter destruction by bringing on a nuclear war.*

- *Unfortunately, we are faced with the ironical fact that while nuclear war seems capable of destroying society, the means to avert it consist in building up the means to wage it. What we must hope for is some measure of arms control.*

- *We must build up a powerful deterrent to war as our first object. Having done that, we must seek to bring about some measure of worldwide disarmament.*

- *Many people put forward proposals for disarmament. But all such proposals depend on a degree of confidence and trust between nations which does not exist.*

- *As Air Forces develop through the jet bomber to the ballistic missile and the satellite, the world balance of power will become progressively more precarious.*

- *I profoundly hope that the full potential of scientific warfare will never be used.*

CHAPTER 9:
THOUGHTS ON PEACE

Montgomery prepares to give a public address during World War II. After the war, Montgomery made tireless efforts to promote world peace. Montgomery's opposition to the Cold War and atomic warfare—as well as his empathic attitude towards student peace activists and members of the "hippie" protest movement—made him unpopular among many former conservative supporters. By contrast, Montgomery believed the "hippie" movement represented a total abandonment of morality and self-discipline, which he did not believe could lead to true freedom or peace. A moderate, independent thinker, Montgomery was caught between the two political extremes of ultraconservatism and ultra-liberalism during the postwar years. He remained true to his convictions at the expense of his reputation. Photo courtesy of Wikimedia Commons.

Montgomery (center right) is accompanied by Winston Churchill and subordinate commanders (circa 1944) as they view a scene of devastation in France during World War II. Montgomery's writings and speeches suggest he was deeply affected by the suffering he witnessed during two world wars, including and especially the Holocaust atrocities he saw at the Bergen-Belsen concentration camp after liberating it. He also strongly condemned Japanese atrocities towards American POWs. Disillusioned by the rivalry of Western and Eastern leaders after World War II, Montgomery tried to build bridges of diplomacy to prevent further bloodshed. Photo courtesy of Wikimedia Commons.

> - *Why are things in the world today in such a mess? Some may say it just happened, and that the reason is inscrutable; I would not agree. Things don't just "happen" in this world of politics and war. They take place because of national policies of lack of them—particularly in respect of wars.*

Montgomery speaks with Dwight D. Eisenhower during a World War II press conference. Montgomery faced a public backlash for criticizing Eisenhower's approach as U.S. president to American foreign policy. While Montgomery's goal was to work towards peace in the postwar era, Eisenhower engaged in Cold War conflicts with the Soviet Union and Communist countries. By contrast, Montgomery visited Russia, China, and South American countries in attempts to build good relations and prevent war. Montgomery's hands-on efforts to promote tolerance of different ideologies and social systems—at the expense of his own career and popularity—are noteworthy in an era of history known for extreme polarization. Montgomery viewed Eisenhower as his "friend and wartime chief" and thought it was possible to remain on good terms with Eisenhower despite their political disagreements. "I am devoted to Ike and would do anything for him," he said, during a 1949 speech in New York. However, Montgomery later wrote that Eisenhower "withdrew his friendship". Photo courtesy of the Museum of Danish Resistance Photo Archives.

- *I think many of the problems that go in the world today are due to the bad leadership on the parts of the people in high places.*

- *Some people seem to think that military actions need be based only on purely military grounds without taking into account their political repercussions. But to think and act thus merely leaves the area of politics in danger. At the top level, can any decision in war be non-political?*

- *Something seems to have gone wrong in the world in which we live. Notwithstanding the progress in civilization—and the longing for peace in the minds of all decent people over the past 2,000 years or more—mankind has not been able to prevent the 20th century from becoming the bloodiest and most turbulent period in recorded history.*

- *I've been through two great world wars, and in neither case have we received the benefits and the peace, which we were led to believe would happen and even promised.*

- *In this close-knit world there is always a danger that war anywhere might spread and involve the whole world, and become total.*

- *Put simply, the issue is this: is it possible for find some way by which states with different ideological doctrines and social systems can live peacefully together without interfering with each other's affairs?*

- *The responsibility of statesmen and politicians is very great. The higher direction of war is in their hands, and they must see to it that they give clear political directives to service chiefs.*

Two sergeants of the American and British armies (Lance Sgt. Brown of the British Eighth Army and Sgt. Randall of the U.S. First Army) were the first in their forces to meet up with each other in April 1943 on the Gabes-Gafsa road in North Africa. Although Montgomery had goodwill towards Americans, he was widely condemned in the U.S. for expressing critical opinions on aspects of American strategy in World War II and postwar American foreign policy. He wrote that although he wished to tour the U.S., he did not because he believed he would be unwelcome. Despite this, Montgomery continued to express many positive opinions about America based on sincere admiration he had developed for the United States and its history over many years, and his cooperation with Americans during World War II. He expressed hope that the political climate would become less reactionary. Photo courtesy of Wikimedia Commons.

- *By the patient exercise of diplomacy—and by the determination to break down suspicion, fear, and even hatred—the mutual relationships between different social systems and ideologies can be made less liable to erupt into war and more consistent with a true meaning of peaceful coexistence.*

Montgomery decorates Russian generals in Berlin in 1945. Although declaring himself morally opposed to Communism, Montgomery made efforts to establish good relations and a sense of mutual understanding with Communist leaders and citizens in Russia, China, South America and other Communist countries. He believed it was impractical to condemn or pressure people for ideological reasons, saying during a 1960s interview: "It so happens that the Eastern part of the world are Communists. Well, it suits them. It's no good saying, "You mustn't be Communists—they are!" He met with many leaders including Joseph Stalin, Nikita Khrushchev, Josip Tito, and Mao Tse-tung during an era when most Western political leaders shunned them. Montgomery wrote he did not believe a person could fairly judge another person without meeting them. During his visits, Montgomery asked hardline questions and attempted to analyze their personalities and beliefs. In doing so, Montgomery was effectively the first Western leader to successfully breach the Iron Curtain during the Cold War. However, he was misunderstood and mocked for his efforts. Due to his courage of conviction and fair-mindedness, the once-popular Field Marshal was heaped with scorn not only in the U.S., but also in his native United Kingdom. In spite of this, Montgomery persisted in his efforts to create goodwill and dialogue among nations. Photo courtesy of Wikimedia Commons.

- *What has got to be achieved is a resolution of the political and ideological differences which divide the world.*

Men of the British Army in Normandy in July 1944. Montgomery never forgot his war experiences or the sacrifices of his troops. Photo courtesy of Wikimedia Commons.

- *The true object in war must be a secure and lasting peace. This will not be brought about if a nation or group of nations go all out for a complete military victory and slam the door on any idea of a negotiated peace—which was what happened in the 1939/45 war, and also in the 1914/18 war.*

- *Until political leaders can find some sensible way of settling international disputes, war will remain with us.*

- *Unfortunately, the time has not yet arrived when we can say that war has been abolished and we have "Peace on Earth." Therefore, political leaders and service chiefs must continue their efforts to make war less horrible.*

- *But there are times in war when men must do a hazardous job, when a position must be held or taken whatever the cost, and when success and a nation's fate depend on the courage, determination, and tenacity of officers and men. Then those who set duty before self give their lives to see the task committed to them through to its completion. They win the day and, in our Christian faith, a higher honor than mortal man can give. It is a free choice. And for the immortal value of that choice, the crosses stand—whatever the religion, faith, or form of worship.*

CHAPTER 10:
THE STUDY OF MILITARY HISTORY

Bernard Montgomery was an enthusiastic military historian. In addition to extensive reading, writing and research on the subject, he frequently visited battlefields, museums and historic military sites whenever possible during his many travels. The numerous places Montgomery visited included crusader castles, ancient warfare sites in Egypt and the Middle East, a Viking exhibit in Norway, and historic battlefields across Europe and the United Kingdom. He also researched the military history of India during his military service in Peshawar, Mumbai (Bombay) and Quetta. Montgomery also wrote some commentary on his personal experiences in war.

–Zita Steele

Montgomery at his writing desk. Photo courtesy of the Museum of Danish Resistance Photo Archives.

- *In my study of war, I've had to study a lot of history. I had to. War is history and history is built up on war.*

- *A vast amount of experience lies buried in the story of past warfare, and commanders could not do without the military historians who uncover it for them. (Nor could they do without us; if there were no generals or admirals to criticize, a lot of them would go out of business!)*

- *Their value is in establishing the facts and in drawing lessons from them (rather than embarking on discussions of what should have been done).*

- *My object was, and always has been, to study the past intelligently, in order to seek guidance for the present and future.*

- *I wanted to discover what was in the great man's mind when he made a major decision. This, surely, was the way to study generalship.*

- *The conduct of war is a life study and if the study has been neglected, a general can expect no success.*

ON SUN TZU:

- *Sun Tzu made some perceptive psychological observations about the relationship between officers and soldiers in the ranks—such as how the general should recognize signs which indicate the state of morale and condition of his troops.*

- *The sentences are terse, sometimes obscure, sometimes deceptively simple, but always full of mature military wisdom—much of which Europeans were not to learn for themselves until the Napoleonic era.*

- *I should like to have talked with Sun Tzu; it would appear that on the subject of the conduct of war we would have much in common, and he understood the human factor.*

Painting of Rear-Admiral Sir Horatio Nelson (1758-1805) by Lemuel Francis Abbot (1799). Image courtesy of Wikimedia Commons.

On Admiral Horatio Nelson:

- *From my earliest days, Nelson has been one of my heroes, and when I myself began to study war it was brought home to me how much that sailor did for England.*

Admiral Horatio Nelson during a sea battle in July 1797 against the Spanish, painting by Richard Westall (1806). Image courtesy of Wikimedia Commons.

- *Great victories on land and sea call forth admiration and respect for the victorious commander-in-chief; but they do not always evoke affection or love.*

- *He knew how to win the hearts of men. He seemed to have a magnetic influence over all who served with him; he led by love and example.*

- *There was nothing he would not do for those who served under him; there was nothing his captains and sailors would not dare for him.*

- *The moment he stepped on board ship some magnetic power radiated from him, a motley collection of men with no common purpose became a band of brothers, and this power radiated far beyond his own ship…*

- *The qualities of leadership in Nelson have always appealed to me enormously.*

Korean Admiral Yi Sun-Sin (1545-1598) was famous for his victories against the Japanese Navy during the Imjin War. Never defeated at sea nor losing a single ship, military historians rank Admiral Yi among the greatest naval commanders in history, such as Admiral Horatio Nelson. Photo courtesy of Wikimedia Commons.

On Admiral Yi Sun-Sin:

- *The Koreans were a seafaring people and they had an outstanding admiral in Yi-sun, who (besides being a strategist, tactician, and leader of exceptional qualities) had a remarkable talent for mechanical invention.*

On Genghis Khan:

- *A great military genius...in the front rank of all soldiers, a grand chef if ever there was one, whose campaigns are models in the art of war.*

On Adolf Hitler:

- *Many regard Hitler as merely an insane political figure. He was, indeed, an evil man, but he was a leader—enterprising and astute...He imparted his evilness to others.*

- *During the years of the Nazi regime in Germany things happened which could not find a parallel in the most debased days of the Roman or Mongol empires, crimes were committed which most people could not imagine—unless they had seen a place like Belsen, which I entered on the day of its liberation with my troops in April 1945. The wholesale liquidation of civilians was unprecedented.*

- *All these things were the responsibility of one evil man—Hitler. Millions starved and died while he and his followers feasted. He parted the wife from her husband, the maid from her lover, the child from its parents. If he had lived, he could never have given back what he had taken from those he had so cruelly wronged—years of life and health and happiness, wives and children, loved ones and friends. If he had 10,000 lives they could not atone, even though each was dragged out to the bitter end in the misery which he meted out to others.*

Zita Steele and Bernard Law Montgomery
On Hannibal and Scipio:

- *It is clear that Hannibal was the better tactician; indeed his tactical genius at Cannae can compare with the conduct of any battle in the history of warfare.*

- *According to Livy, he proposed to Scipio that the two of them should meet between the armies and talk things over. I find it hard to imagine such a thing happening in the 20th century; if Rommel had asked me to meet him between our lines before Alamein for a discussion on the situation, I would have declined—although I would have been intensely interested to meet my famous opponent, which I never did.*

- *If Rommel had said to me at Alamein, 'Let us meet and have a talk and see if we can't do something about it,' I would have said, 'No.' Although I would have been very pleased to see my famous opponent. No. This is war. I'm going to defeat you—smash you in battle.*

- *Where Scipio was undoubtedly superior to Hannibal was in strategy. It was this which in the end mattered most, and which marks out Scipio as one of the great captains in history.*

On Julius Caesar:

- *For a general to have achieved such total overall success, Caesar is extraordinarily open to criticism.*

- *He threw away the advantage of his surprise crossing into Macedonia by informing Pompey of his presence, and by wasting time in Egypt and Pontus, he allowed the Pompeians to rally and reorganize in Africa.*

- *As a tactician Caesar showed no originality. He neglected cavalry and fought all his various enemies with a three-line legion which was basically extremely traditional. He was, however, by far the greatest Roman infantry commander; and he raised the legion to its highest point.*

- *His only concern was power, and to secure it he was completely ruthless and amoral...It is open to question whether towards the end of his life he was altogether sane. Caesar is certainly the most disappointing of great conquerors.*

On Saladin:

- *Saladin was a most able ruler, a devout Muslim, and a sound strategist.*

On Flavius Belisarius:

- *He [Belisarius] is the classic example of a loyal and capable soldier compelled by a second-rate political chief to pursue an unrealistic strategical objective.*

A Byzantine military general, Belisarius (circa 505 to 565 A.D.) was instrumental in the reconquest of much of the Mediterranean territory that had been part of the Western Roman Empire. Photo courtesy of Wikimedia Commons.

Napoleon crossing the alps, painting by Jacques-Louis David (circa 1801). Image courtesy of Wikimedia Commons.

ON NAPOLEON BONAPARTE:

• *Many judgments have been passed on Napoleon and his deeds. It has always seemed to me that he was too ambitious. He was determined to be known as the greatest general ever, and this ambition drove him on to final defeat.*

• *But one thing can be said for certain—his victories have not been surpassed, and so long as there are soldiers he will be remembered as one of the greatest of the captains.*

—Bernard Montgomery's Art of War—

MacArthur (left) and Nimitz discuss Pacific war strategy in Brisbane, Australia in March 1944 at MacArthur's headquarters. Photograph courtesy the National Archives.

On General Douglas MacArthur and Admiral Chester Nimitz::

- *MacArthur and Nimitz (both of whom I knew) fought brilliant campaigns in their war against the Japanese in the Far East. It was entirely an American war; they fought a novel war by novel means and saw it through to complete success.*

- *I am glad I was privileged to know Admiral Nimitz—a very great sailor.*

Union and Confederate veterans of the American Civil War shake hands at a reunion commemorating the anniversary of the Battle of Gettysburg in 1913. Photograph courtesy of the Library of Congress. Montgomery was an enthusiastic Civil War researcher. He was a great admirer of U.S. President Abraham Lincoln, to whom he devoted an entire chapter in one of his books. Montgomery praised Lincoln for ending slavery in the U.S. and always visited the Lincoln Memorial on trips to Washington, D.C. He described men on both sides of the war as "magnificent natural soldiers." Montgomery also visited the Gettysburg battlefield, including the cemetery and site of Lincoln's Gettysburg Address.

On the American Civil War:

> • *My own study of the war has revealed that whereas the generalship was not too good, what might be called 'the soldiership' was first class—the men in the ranks on both sides being magnificent natural soldiers, ready and willing to fight for the cause in which they believed and, if necessary, to die for that cause. My final word on the American Civil War is that it well repays study by soldiers of the present day.*

On World War I:

- *The only impressive results in that theatre were the casualties, and these had a profound influence on my military thinking.*

- *As it seems to me, nobody tried to prevent the war.*

On World War II:

- *The war which thus engulfed the world was totally different from the conflict of 1914/18...I who fought in both can truly say that it would be impossible to find two wars against the same enemy which were more different.*

- *Hitler made it clear in Mein Kampf that peace was merely a period of preparation for total war, and the Nazi state became a war machine tuned for action. The former corporal himself took over the leadership of the state and of the high command, and those Germans who did not accept guns as well as butter were ruthlessly dealt with.*

- *During the war, crimes were committed by the Germans and Japanese which have, I think, no parallel in scale and wickedness in history.*

- *This was an immensely complicated war...The scene is constantly changing and covers the entire planet except for the land mass of the American continents.*

- *It was, I suppose, the greatest tragedy in the history of mankind.*

- *The amount of human suffering was beyond all belief.*

MONTGOMERY'S READING LIST

Montgomery was an avid reader and studied the works of many writers on a wide variety of subjects, including military history. The following is a short list of some war-related writings he recommended or referred to favorably in his published works.

—Zita Steele

Books

"The Arthashastra" by Kautilya (ancient Indian Sanskrit composed between the 2nd and 3rd century B.C.).

"The Defense of Duffer's Drift" (1904) and *"The Green Curve"* (1914) by Major-General Sir Ernest Dunlop Swinton.

"The Art of Warfare in Biblical Lands—in the light of Archaeological Study" by Yigael Yadin (1963).

Yigael Yadin, Israeli soldier, archeologist and politician. Photo courtesy of IDF Spokesperson's Unit/Wikimedia Commons.

Sir Ernest Dunlop Swinton, British Army office and military writer who influenced the development and use of the tank by the British during World War II. Photos courtesy of Wikimedia Commons.

"*The Origins of the Second World War*" by A.J.P. Taylor (1961).

"*The Memoirs of Captain Liddell Hart: Volumes I and II*" by Sir Basil Henry Liddell Hart (1965).

"*Selected Military Writings*" by Mao Tse-tung (1967).

"*The Science of War— a Collection of Essays and Lectures*" by Colonel George Frances Robert Henderson (1905).

Colonel George Frances Robert (G.F.R.) Henderson was a noted British officer, military writer, lecturer and instructor. He taught Tactics, Military Administration and Law at Sandhurst and was a professor of Military Art and History. He was a scholar on the American Civil War and visited numerous battlefields. His military service took him to India, Egypt and South Africa. During the Second Boer War, "Henderson, always an ardent advocate for mystifying and misleading the enemy, was especially active, and reveled in the deceits he practiced. He sent out fictitious telegrams to commanders in clear, and then on one excuse or another countermanded them in cipher; circulated false orders implying a concentration of troops at Colesberg ... gave confidential tips to people eager for news whom he knew would at once divulge them ... On the whole it is probable that no military plan was ever kept better concealed from friend or foe." Photo courtesy of Wikimedia Commons.

Epic Poems on the Theme of War

Illustration in "The Iliad of Homer, Translated by Mr. Pope" written by Homer and translated by Alexander Pope (1720), courtesy of Wikimedia Commons.

"*The Iliad*" by Homer (epic Greek poem set in the Trojan War, 8th century B.C.).

"*The Mahabharata*" (Sanskrit tales of ancient India, circa 4th century).

"*The Song of Roland*" (epic poem based on the Battle of Roncevaux Pass, circa 1040 to 1115 A.D.).

Eight phases of The Song of Roland by Simon Marmion, in Grandes Chroniques de France (15th century), courtesy of Wikimedia Commons.

Siegfried hands the drinking-horn back to Gutrune in "Siegfried & The Twilight of the Gods" by Richard Wagner, illustration by Arthur Rackham (1924); courtesy of Wikimedia Commons.

"*The Nibelungenlied*" (heroic epic poem translated as "The Song of the Nibelungs", circa 1200 A.D.).

About the Author:
Zita Steele

Photo by Noël-Marie Fletcher.

Zita Ballinger Fletcher (also known by the pen name Zita Steele) is a journalist, author, and award-winning military history writer. She writes fiction and nonfiction books, and has published more than 10 works. With a background in art, she designs and illustrates her published work. She also produces videos and creates multimedia content.

Zita is the author of the first published collection of Field Marshal Erwin Rommel's wartime photography, a groundbreaking military history book series. Her December 2019 military history article on Field Marshal Bernard Law Montgomery *("Monty Crosses the Rhine")* won a 2nd Place award in the National Federation of Press Women's 2020 nationwide writing contest.

Zita is a member of the Military Writers Society of America, the British Military Historical Society, and the Friends of the Fusilier Museum Warwick (conserving the history of the Royal Warwickshire Regiment) in the U.K. Her areas of interest include World War II British and Commonwealth history and German Resistance.

Zita is fluent in German and studies Russian. She attended the Honors College at the University of South Florida and graduated Magna Cum Laude with an Honors Degree in Social Sciences. Her writings have been published by: *Military History* magazine, *World War II Quarterly* journal, *World War II History* magazine, The Max Kade Institute for German-American Studies, *The Abraham Lincoln Association's For the People Newsletter, North Irish Roots* (UK), and the *Journals of the Gloucestershire* and *Lincolnshire Family History Societies* (UK). She is also a photographer. She has done research on location at many historic war-related sites in Europe and the U.S., including museums, institutions and former battlefields.

Zita's "Author" interview with the *Military Writers Society of America* (MWSA), Jan. 19, 2020.

MWSA: Would you recommend MWSA membership to other authors?

Zita Fletcher: *Definitely. MWSA provides so many resources for professional development and opportunities to learn from and connect with a wonderful community of fellow authors. I am happy to be part of the MWSA writing community and encourage all prospective authors to join. No matter what your experience level, you will find opportunities to learn, engage with others and share your story.*

MWSA: Please tell us a little about your writing background and philosophy.

Zita Fletcher: *I've been writing stories for as long as I can remember. I excelled at writing in school and loved storytelling. My earliest ambition was to write books.*

Besides writing, history and social studies were my strongest subjects in school. They continue to be my strongest subjects.

In college, I majored in Social Sciences with concentrations in International Studies and Criminology. Within my discipline, I focused on criminal profiling and psychology. I earned many academic distinctions. After graduation, I strongly considered pursuing a career in Forensic Psychology. I ultimately decided to become a professional writer and am happy with my choice.

As a journalist, I've enjoyed being able to share the stories and thoughts of many of our country's veterans including a Marine medic who fought in the Battle of Guam, a Korean War veteran, wounded veterans who participated in the Warriors to Lourdes journey, a Medal of Honor recipient and others.

As an American, I'm a strong believer in intellectual freedom and the human right to question. Freely exchanging opinions and firsthand learning are opportunities for discovery and enlightenment. Throughout my life I have rejected attempts by others to control or dictate what I think or believe. I became a political independent in college and remain so. I reject labels. I use the term "free thinker" to emphasize my philosophy of independently determining my beliefs and values.

MWSA: Why do you write under a pen name as an author?

Zita Fletcher: *As a creative person, I like to have the flexibility of writing under a creative name to express myself. I use the pen name Zita "Steele." The name "Steele" is a wordplay on steel metal. According to the Chinese zodiac and Five Element system, my element is Metal. I do not believe in astrology or horoscopes; I simply find this very cool.*

MWSA: When did you become interested in military matters?

Zita Fletcher: *My interest in war, soldiers and generals dates from an early age. As a child, I liked to play with toy knights instead of dolls. Also growing up in a Catholic house, I had a strong attraction to muscular male saints (often depicted with weapons) including St. Michael, St. George, and St. Sebastian. When I was 10, people were astonished to learn that my favorite movie was Ridley Scott's "Gladiator." I also amazed others when I bought Sun Tzu's "Art of War" during a bookstore trip at about age 12.*

I was one of the only girls I knew who liked shooting games and war movies. On trips to museums, I was fascinated by guns, armor, swords and spears. In high school I was a voracious reader of ancient Roman war histories and keenly interested in samurai. One of my favorite TV shows was "Human Weapon." I liked to watch the Military Channel, which is now called the American Heroes Channel.

I think my interests surprised people who did not expect a feminine young lady to have such a strong attraction to warriors and military science.

My genetic makeup is English, Irish, German, and Spanish—all ethnicities known for having great fighting spirit and strength of character.

I was also born in the Chinese zodiac Year of the Horse. While I do not believe in astrology, some traits associated with the Horse accurately describe my personality.

I was raised in a single-parent household; given my strong-willed nature, things could have been difficult. Thankfully my mom appreciated my free spirit and allowed me to be very independent growing up.

In addition to military history, I've also had a lifelong interest in martial arts and sports. My favorite sports include mixed martial arts and fencing (foil); I started foil fencing at age 17, and still love it.

MWSA: What do you like about military history?

Zita Fletcher: *One of the main reasons I love military history is because it is filled with courage, leadership and self-sacrifice. I like to* *learn about feats of bravery and strength. There is also a lot of warmth and humor among soldiers. I admire stories of brotherhood and great commanders who bonded with their troops.*

I've always been very curious about the international landscape—I love travel and foreign languages. This has led me to have keen interest in military history and theories from other countries.

MWSA: Do you come from a military family?

Zita Fletcher: *My family history includes military tradition.*

My great-grandfather Edward W. Arnold was a U.S. Marine Corps instructor during World War I. My grandfather Ray A. Fletcher Sr. served as a medic in the U.S. Army Air Forces during World War II and as a Lieutenant Colonel in the Civil Air Patrol during the Korean War.

My direct ancestor, Morris Gibbons, was a "bushwhacker" guerilla chief who led raids under Col. Joe Porter in Missouri during the Civil War. My other ancestors include a Southern cavalryman in Hampton's Legion, several Revolutionary War-era militia captains, a British Royal Navy privateer, a Swiss-German mercenary who fought for the Duke of Marlborough and Spanish conquistadors.

MWSA: Why do you speak German?

Zita Fletcher: *I studied German for many years as part of a personal journey to get in touch with my heritage. Aside from the rewarding experience of connecting with other people, this skill has also been priceless regarding military history. I've been able to do research at many German-language institutions in Europe, including the German Historical Museum in Berlin, the Jewish Museum of Berlin, the Nuremberg Trials Memorial, the Munich City Museum, the Museum of Military History in Vienna, and many other places.*

I find there is a void when it comes to sharing German history and perspectives with English readers—things get lost in translation, or just simply lost. I use my German language skills to research and share knowledge through my writing.

MWSA: Where did your interest in British military history come from?

Zita Fletcher: The first time I encountered the British military was through research on World War II North Africa for a project that started when I was 15. The desert war has remained a major interest for me ever since.

I was impressed with the brave soldiers of the British Eighth Army. I learned so much from the compelling stories of these heroic men from England and the Commonwealth countries. I was also impressed with the Eighth Army's leader, Field Marshal Bernard L. Montgomery. This led me to develop interests in other aspects of the British military and its rich history.

I also personally relate to and admire British people and culture given my strong British heritage (Northern Irish + English). British history is a great source of inspiration to me.

MWSA: Why did you decide to write a book series on Field Marshal Erwin Rommel?

Zita Fletcher: Rommel is a controversial figure—but that is why he is interesting. We cannot ignore controversy if we want to learn from the past. Studying Rommel gives insights into the experiences of a German soldier who began his career in an era of turmoil, rose in the ranks and was ultimately killed by the Nazi system. There are many aspects of his story that are worth examining.

What first got my attention regarding Rommel was his photography collection, which I found fascinating. I created my book series because I wanted to share my discoveries.

Rommel is not the only general I find interesting. Other military leaders I've enjoyed studying include Hannibal, Yi Sun Shin, George Washington and T.E. Lawrence—and of course my favorite commander, Field Marshal Bernard Montgomery.

www.ingramcontent.com/pod-product-compliance
Lightning Source LLC
Chambersburg PA
CBHW032111090426
42743CB00007B/315